P9-AQJ-882

QUANTUM KINEMATICS AND DYNAMICS

Frontiers in Physics

A Lecture Note and Reprint Series

DAVID PINES, *Editor*

N. Bloembergen NUCLEAR MAGNETIC RELAXATION
A Reprint Volume

Geoffrey F. Chew S-MATRIX THEORY OF STRONG INTERACTIONS
A Lecture Note and Reprint Volume

R. P. Feynman QUANTUM ELECTRODYNAMICS
A Lecture Note and Reprint Volume

R. P. Feynman THE THEORY OF FUNDAMENTAL PROCESSES
A Lecture Note Volume

Hans Frauenfelder THE MOSSBAUER EFFECT
A Collection of Reprints with an Introduction

David Pines THE MANY-BODY PROBLEM
A Lecture Note and Reprint Volume

L. Van Hove, N. M. Hugenholtz, and L. P. Howland
PROBLEMS IN THE QUANTUM THEORY OF MANY-PARTICLE SYSTEMS

Julian Schwinger QUANTUM KINEMATICS AND DYNAMICS

QUANTUM KINEMATICS AND DYNAMICS

JULIAN SCHWINGER

Harvard University

W. A. BENJAMIN, INC.

New York 1970

QUANTUM KINEMATICS AND DYNAMICS

Standard Book Numbers: 8053-8510-X (C)
8053-8511-8 (P)
Library of Congress Catalog Card Number 72-135371
Manufactured in the United States of America
12345 R43210

W. A. BENJAMIN, INC.
New York, New York 10016

Foreword

Early in 1955 I began to write an article on the Quantum Theory of Fields. The introduction contained this description of its plan. "In part A of this article a general scheme of quantum kinematics and dynamics is developed within the nonrelativistic framework appropriate to systems with a finite number of dynamical varibles. Apart from specific physical consequences of the relativistic invariance requirement, the extension to fields in part B introduces relatively little that is novel, which permits the major mathematical features of the theory of fields to be discussed in the context of more elementary physical systems."

A preliminary and incomplete version of part A was used as the basis of lectures delivered in July, 1955 at the Les Houches Summer School of Theoretical Physics. Work on part A ceased later that year and part B was never begun. Several years after, I used some of the material in a series of notes published in the Proceedings of the National Academy of Sciences. And there the matter rested until, quite recently, Robert Kohler (State University College at Buffalo) reminded me of the continuing utility of the Les Houches notes and suggested their publication. He also volunteered to assist in this process. Here is the result. The main text is the original and still incomplete 1955 manuscript, modified only by the addition of subheadings. To it is appended excerpts from the Proc. Nat. Acad. of Sciences articles that supplement the text, together with two papers that illustrate and further develop its methods.

Belmont, Massachusetts
1969

Julian Schwinger

Contents

CHAPTER ONE
THE ALGEBRA OF MEASUREMENT

The classical theory of measurement is built upon the conception of an interaction between the system of interest and the measuring apparatus that can be made arbitrarily small, or at least precisely compensated, so that one can speak meaningfully of an idealized measurement that disturbs no property of the system. But it is characteristic of atomic phenomena that the interaction between system and instrument is not arbitarily small. Nor can the disturbance produced by the interaction be compensated precisely since to some extent it is uncontrollable and unpredictable. Accordingly, a measurement on one property can produce unavoidable changes in the value previously assigned to another property, and it is without meaning to speak of a microscopic system possessing precise values for all its attributes. This contradicts the classical representation of all physical quantities by numbers. The laws of atomic physics must be expressed, therefore, in a nonclassical mathematical language that constitutes a symbolic expression of the properties of microscopic measurement.

1.1 MEASUREMENT SYMBOLS

We shall develop the outlines of this mathematical structure by discussing simplified physical systems which are such that any physical quantity A assumes only a finite number of distinct values, $a', \ldots a^n$. In the most elementary type of measurement, an ensemble of independent similar systems is sorted by the apparatus into subensembles, distinguished by definite values of the physical quantity being measured. Let $M(a')$ symbolize the selective measurement that accepts systems possessing the value a' of property A and rejects all others. We define the addition of such symbols to signify less specific selective measurements that produce a subensemble associated with any of the values in the summation, none of these being distinguished by the measurement.

The multiplication of the measurement symbols represents the successive performance of measurements (read from right to left). It follows from the physical meaning of these operations that addition is commutative and associative, while multiplication is associative. With 1 and 0 symbolizing the measurements that, respectively, accept and reject all systems, the properties of the elementary selective measurements are expressed by

$$M(a')M(a') = M(a') \tag{1.1}$$

$$M(a')M(a'') = 0 \quad , \quad a' \neq a'' \tag{1.2}$$

$$\sum_{a'} M(a') = 1 \,. \tag{1.3}$$

Indeed, the measurement symbolized by $M(a')$ accepts every system produced by $M(a')$ and rejects every system produced by $M(a'')$, $a'' \neq a'$, while a selective measurement that does not distinguish any of the possible values of a' is the measurement that accepts all systems.

According to the significance of the measurements denoted as 1 and 0, these symbols have the algebraic properties

$$\begin{aligned} 1\,1 &= 1 \quad , \quad 0\,0 = 0 \\ 1\,0 &= 0\,1 = 0 \\ 1 + 0 &= 1 \,, \end{aligned} \tag{1.4}$$

and

$$\begin{aligned} 1M(a') &= M(a')1 = M(a') \,, \\ 0M(a') &= M(a')0 = 0 \\ M(a') + 0 &= M(a') \,, \end{aligned} \tag{1.5}$$

which justifies the notation. The various properties of 0, $M(a')$ and 1 are consistent, provided multiplication is distributive. Thus,

$$\sum_{a''} M(a')M(a'') = M(a') = M(a')1$$
$$= M(a')\sum_{a''} M(a'') . \tag{1.6}$$

The introduction of the numbers 1 and 0 as multipliers, with evident definitions, permits the multiplication laws of measurement symbols to be combined in the single statement

$$M(a')M(a'') = \delta(a',a'')M(a') , \tag{1.7}$$

where

$$\delta(a',a'') = \begin{cases} 1 , & a' = a'' \\ 0 , & a' \neq a'' . \end{cases} \tag{1.8}$$

1.2 COMPATIBLE PROPERTIES. DEFINITION OF STATE

Two physical quantities A_1 and A_2 are said to be compatible when the measurement of one does not destroy the knowledge gained by prior measurement of the other. The selective measurements $M(a_1')$ and $M(a_2')$, performed in either order, produce an ensemble of systems for which one can simultaneously assign the values a_1' to A_1 and a_2' to A_2 . The symbol of this compound measurement is

$$M(a_1'a_2') = M(a_1')M(a_2') = M(a_2')M(a_1') \quad . \tag{1.9}$$

By a complete set of compatible physical quantities, $A_1, \ldots A_k$, we mean that every pair of these quantities is compatible and that no other quantities exist, apart from functions of the set A , that are compatible with every member of this set. The measurement symbol

$$M(a') = \prod_{i=1}^{k} M(a_i') \tag{1.10}$$

then describes a complete measurement, which is such that the systems chosen possess definite values for the maximum number of attributes; any attempt to determine the value of still another independent physical quantity will produce uncontrollable changes in one or more of the previously assigned values. Thus the optimum state of knowledge concerning a given system is realized by subjecting it to a complete selective measurement. The systems admitted by the complete measurement $M(a')$ are said to be in the state a' . The symbolic properties of complete measurements are also given by (1.1), (1.2) and (1.3).

1.3 MEASUREMENTS THAT CHANGE THE STATE

A more general type of measurement incorporates a disturbance that produces a change of state. The symbol $M(a', a'')$ indicates a selective measurement in which systems are accepted only in the state a'' and emerge in the state a'. The measurement process $M(a')$ is the special case for which no change of state occurs,

$$M(a') = M(a', a') \quad . \tag{1.11}$$

The properties of successive measurements of the type $M(a', a'')$ are symbolized by

$$M(a', a'')M(a''', a^{iv}) = \delta(a'', a''') M(a', a^{iv}) \quad , \tag{1.12}$$

for, if $a'' \neq a'''$, the second stage of the compound apparatus accepts none of the systems that emerge from the first stage, while if $a'' = a'''$, all such systems enter the second stage and the compound measurement serves to select systems in the state a^{iv} and produce them in the state a'. Note that if the two stages are reversed, we have

$$M(a''', a^{\text{iv}})M(a', a'') = \delta(a', a^{\text{iv}})M(a''', a''), \qquad (1.13)$$

which differs in general from (1.12). Hence the multiplication of measurement symbols is noncommutative.

The physical quantities contained in one complete set A do not comprise the totality of physical attributes of the system. One can form other complete sets, $B, C, \ldots,$ which are mutually incompatible, and for each choice of non-interfering physical characteristics there is a set of selective measurements referring to systems in the appropriate states, $M(b', b''), M(c', c''), \ldots$. The most general selective measurement involves two incompatible sets of properties. We symbolize by $M(a', b')$ the measurement process that rejects all impinging systems except those in the state b', and permits only systems in the state a' to emerge from the apparatus. The compound measurement $M(a', b')M(c', d')$ serves to select systems in the state d' and produce them in the state a', which is a selective measurement of the type $M(a', d')$. But, in addition, the first stage supplies systems in the state c' while the second stage accepts only systems in the state b'. The examples of compound measurements that we have already considered involve the passage of all systems or no systems between the two stages, as represented by the multiplicative numbers 1 and 0. More generally, measurements of properties B, performed on a system in a state c' that refers to properties incompatible with B, will yield a statistical distribution of the possible values. Hence, only a determinate fraction of the systems emerging from the first stage will be accepted by the second stage. We express this by the general multiplication law

$$M(a', b')M(c', d') = \langle b'|c'\rangle M(a', d'), \tag{1.14}$$

where $\langle b'|c'\rangle$ is a number characterizing the statistical relation between the states b' and c'. In particular,

$$\langle a'|a''\rangle = \delta(a', a'') \quad . \tag{1.15}$$

1.4 TRANSFORMATION FUNCTIONS

Special examples of (1.14) are

$$M(a')M(b', c') = \langle a'|b'\rangle \, M(a', c') \tag{1.16}$$

and

$$M(a', b')M(c') = \langle b'|c'\rangle \, M(a', c'). \tag{1.17}$$

We infer from the fundamental measurement symbol property (1.3) that

$$\begin{aligned} \sum_{a'} \langle a'|b'\rangle M(a', c') &= \sum_{a'} M(a')M(b', c') \\ &= M(b', c') \end{aligned} \tag{1.18}$$

and similarly

$$\sum_{c'} \langle b'|c'\rangle \, M(a', c') = M(a', b'), \tag{1.19}$$

which shows that measurement symbols of one type can be expressed as a linear combination of the measurement symbols of another type. The general relation is

$$\begin{aligned} M(c', d') &= \sum_{a'b'} M(a')M(c', d')M(b') \\ &= \sum_{a'b'} \langle a'|c'\rangle \langle d'|b'\rangle M(a', b'). \end{aligned} \tag{1.20}$$

From its role in effecting such connections, the

totality of numbers $\langle a'|b'\rangle$ is called the transformation function relating the a - and the b-descriptions, where the phrase "a -description" signifies the description of a system in terms of the states produced by selective measurements of the complete set of compatible physical quantities A.

A fundamental composition property of transformation functions is obtained on comparing

$$\sum_{b'} M(a')M(b')M(c') = \sum_{b'} \langle a'|b'\rangle\langle b'|c'\rangle \, M(a', c') \tag{1.21}$$

with

$$M(a')\left(\sum_{b'} M(b')\right)M(c') = M(a')M(c') = \langle a'|c'\rangle \, M(a', c'), \tag{1.22}$$

namely

$$\sum_{b'} \langle a'|b'\rangle \langle b'|c'\rangle = \langle a'|c'\rangle \, . \tag{1.23}$$

On identifying the a - and c -descriptions this becomes

$$\sum_{b'} \langle a'|b'\rangle\langle b'|a''\rangle = \delta(a', a'') \tag{1.24}$$

and similarly

$$\sum_{a'} \langle b'|a'\rangle \langle a'|b''\rangle = \delta(b', b'') . \qquad (1.25)$$

As a consequence, we observe that

$$\sum_{a'} \sum_{b'} \langle a'|b'\rangle \langle b'|a'\rangle = \sum_{a'} 1$$
$$= \sum_{b'} \sum_{a'} \langle b'|a'\rangle \langle a'|b'\rangle = \sum_{b'} 1, \qquad (1.26)$$

which means that N, the total number of states obtained in a complete measurement, is independent of the particular choice of compatible physical quantities that are measured. Hence the total number of measurement symbols of any specified type is N^2. Arbitrary numerical multiples of measurement symbols in additive combination thus form the elements of a linear algebra of dimensionality N^2 - the algebra of measurement. The elements of the measurement algebra are called operators.

1.5 THE TRACE

The number $\langle a'|b'\rangle$ can be regarded as a linear numerical function of the operator $M(b', a')$. We call this linear correspondence between operators and numbers the trace,

$$\langle a'|b'\rangle = \operatorname{tr} M(b', a'), \tag{1.27}$$

and observe from the general linear relation (1.20) that

$$\begin{aligned} \operatorname{tr} M(c', d') &= \sum_{a'b'} \langle a'|c'\rangle \langle d'|b'\rangle \operatorname{tr} M(a', b') \\ &= \sum_{a'b'} \langle d'|b'\rangle \langle b'|a'\rangle \langle a'|c'\rangle \\ &= \langle d'|c'\rangle \quad , \end{aligned} \tag{1.28}$$

which verifies the consistency of the definition (1.27). In particular,

$$\begin{aligned} \operatorname{tr} M(a', a'') &= \delta(a', a'') \\ \operatorname{tr} M(a') &= 1 \ . \end{aligned} \tag{1.29}$$

The trace of a measurement symbol product is

$$\begin{aligned} \operatorname{tr} M(a', b')M(c', d') &= \langle b'|c'\rangle \operatorname{tr} M(a', d') \\ &= \langle b'|c'\rangle \langle d'|a'\rangle \ , \end{aligned} \tag{1.30}$$

which can be compared with

$$\begin{aligned} \operatorname{tr} M(c', d')M(a', b') &= \langle d'|a'\rangle \operatorname{tr} M(c', b') \\ &= \langle d'|a'\rangle \langle b'|c'\rangle \ . \end{aligned} \tag{1.31}$$

Hence, despite the noncommutativity of multiplication, the trace of a product of two factors is independent of the multiplication order. This applies to any two elements X, Y, of the measurement algebra,

$$\mathrm{tr}\, XY = \mathrm{tr}\, YX \quad . \tag{1.32}$$

A special example of (1.30) is

$$\mathrm{tr}\, M(a')M(b') = \langle a'|b'\rangle\langle b'|a'\rangle \quad . \tag{1.33}$$

1.6 STATISTICAL INTERPRETATION

It should be observed that the general multiplication law and the definition of the trace are preserved if we make the substitutions

$$\begin{aligned} M(a', b') &\to \lambda(a')^{-1}\, M(a', b')\lambda(b') \\ \langle a'|b'\rangle &\to \lambda(a')\, \langle a'|b'\rangle\, \lambda(b')^{-1} \quad , \end{aligned} \tag{1.34}$$

where the numbers $\lambda(a')$ and $\lambda(b')$ can be given arbitrary non-zero values. The elementary measurement symbols $M(a')$ and the transformation function $\langle a'|a''\rangle$ are left unaltered. In view of this arbitrariness, a transformation function $\langle a'|b'\rangle$ cannot, of itself, possess a direct physical interpretation but must enter in some combination that remains invariant under the substitution (1.34).

The appropriate basis for the statistical interpretation of the transformation function can be inferred by a consideration of the sequence of selective measurements $M(b')M(a')M(b')$, which differs from $M(b')$ in virtue of the disturbance attendant upon the intermediate A-measurement. Only a fraction of the systems selected in the initial B-measurement is transmitted through the complete apparatus. Correspondingly, we have the symbolic equation

$$M(b')M(a')M(b') = p(a', b')M(b') \quad , \tag{1.35}$$

where the number

$$p(a', b') = \langle a'|b'\rangle\langle b'|a'\rangle \tag{1.36}$$

is invariant under the transformation (1.34). If we perform an A-measurement that does not distinguish between two (or more) states, there is a related additivity of the numbers $p(a', b')$,

$$\begin{aligned} &M(b')\left(M(a') + M(a'')\right) M(b') \\ &= \left(p(a', b') + p(a'', b')\right) M(b') \quad , \end{aligned} \tag{1.37}$$

and, for the A-measurement that does not distinguish among any of the states, there appears

$$M(b')\left(\sum_{a'} M(a')\right) M(b') = M(b'), \tag{1.38}$$

whence

$$\sum_{a'} p(a', b') = 1 \quad . \tag{1.39}$$

These properties qualify p(a', b') for the role of the probability that one observes the state a' in a measurement performed on a system known to be in the state b'. But a probability is a real, non-negative number. Hence we shall impose an admissible restriction on the numbers appearing in the measurement algebra, by requiring that $\langle a'|b'\rangle$ and $\langle b'|a'\rangle$ form a pair of complex conjugate numbers

$$\langle b'|a'\rangle = \langle a'|b'\rangle^* \quad , \tag{1.40}$$

for then

$$p(a', b') = |\langle a'|b'\rangle|^2 \geq 0 \quad . \tag{1.41}$$

To maintain the complex conjugate relation (1.40), the numbers $\lambda(a')$ of (1.34) must obey

$$\lambda(a')^* = \lambda(a')^{-1} \quad , \tag{1.42}$$

and therefore have the form

$$\lambda(a') = e^{i\varphi(a')} \tag{1.43}$$

in which the phases $\varphi(a')$ can assume arbitrary real values.

1.7 THE ADJOINT

Another satisfactory aspect of the probability formula (1.36) is the symmetry property

$$p(a', b') = p(b', a') \ . \tag{1.44}$$

Let us recall the arbitrary convention that accompanies the interpretation of the measurement symbols and their products - the order of events is read from right to left (sinistrally). But any measurement symbol equation is equally valid if interpreted in the opposite sense (dextrally). and no physical result should depend upon which convention is employed. On introducing the dextral interpretation, $\langle a'|b'\rangle$ acquires the meaning possessed by $\langle b'|a'\rangle$ with the sinistral convention. We conclude that the probability connecting states a' and b' in given sequence must be constructed symmetrically from $\langle a'|b'\rangle$ and $\langle b'|a'\rangle$. The introduction of the opposite convention for measurement symbols will be termed the adjoint operation, and is indicated by $\dagger$. Thus,

$$M(a', b')^\dagger = M(b', a') \tag{1.45}$$

and

$$M(a', a'')^\dagger = M(a'', a') \quad . \tag{1.46}$$

In particular,

$$M(a')^\dagger = M(a') \quad , \tag{1.47}$$

which characterizes $M(a')$ as a self-adjoint or Hermitian operator. For measurement symbol products we have

$$\begin{aligned}\big(M(a', b')M(c', d')\big)^\dagger &= M(d', c')M(b', a') \\ &= M(c', d')^\dagger M(a', b')^\dagger \quad ,\end{aligned} \tag{1.48}$$

or equivalently,

$$\begin{aligned}\big(\langle b'|c'\rangle M(a', d')\big)^\dagger &= \langle c'|b'\rangle M(d', a') \\ &= \langle b'|c'\rangle^* M(a', d')^\dagger \quad .\end{aligned} \tag{1.49}$$

The significance of addition is uninfluenced by the adjoint procedure, which permits us to extend these properties to all elements of the measurement algebra:

$$(X+Y)^\dagger = X^\dagger + Y^\dagger \ , \quad (XY)^\dagger = Y^\dagger X^\dagger \ , \quad (\lambda X)^\dagger = \lambda^* X^\dagger \ , \tag{1.50}$$

in which λ is an arbitrary number.

1.8 COMPLEX CONJUGATE ALGEBRA

The use of complex numbers in the measurement algebra implies the existence of a dual algebra in which all numbers are replaced by the complex conjugate numbers. No physical result can depend upon which algebra is employed. If the operators of the dual algebra are written X^*, the correspondence between the two algebras is governed by the laws

$$(X+Y)^* = X^* + Y^* , \quad (XY)^* = X^*Y^* , \quad (\lambda X)^* = \lambda^* X^* . \tag{1.51}$$

The formation of the adjoint within the complex conjugate algebra is called transposition,

$$X^T = X^{*\dagger} = X^{\dagger *} . \tag{1.52}$$

It has the algebraic properties

$$(X+Y)^T = X^T + Y^T , \quad (XY)^T = Y^T X^T , \quad (\lambda X)^T = \lambda X^T . \tag{1.53}$$

1.9 MATRICES

The measurement symbols of a given description provide a basis for the representation of an arbitrary operator by N^2 numbers, and the abstract properties of operators are realized by the combinatorial laws of these arrays of numbers, which are those of matrices. Thus

$$X = \sum_{a'a''} \langle a'|X|a''\rangle M(a', a'') \tag{1.54}$$

defines the matrix of X in the a-description or a-representation, and the product

$$\begin{aligned} XY &= \sum \langle a'|X|a''\rangle M(a', a'') \sum \langle a^{iv}|Y|a'''\rangle M(a^{iv}, a''') \\ &= \sum \langle a'|X|a''\rangle\, \delta(a'', a^{iv})\, \langle a^{iv}|Y|a'''\rangle M(a', a''') \end{aligned} \tag{1.55}$$

shows that

$$\langle a'|XY|a'''\rangle = \sum_{a''} \langle a'|X|a''\rangle\langle a''|Y|a'''\rangle \quad . \tag{1.56}$$

The elements of the matrix that represents X can be expressed as

$$\langle a'|X|a''\rangle = \text{tr}\, XM(a'', a') \quad , \tag{1.57}$$

and in particular

$$\langle a'|X|a'\rangle = \text{tr}\, XM(a') \quad . \tag{1.58}$$

The sum of the diagonal elements of the matrix is the trace of the operator. The corresponding basis in the dual algebra is $M(a', a'')^*$, and the matrices that represent X^* and X^T are the complex conjugate and transpose, respectively, of the matrix representing X . The operator $X^\dagger = X^{T*}$, an element of the same algebra as X , is represented by the transposed, complex conjugate, or adjoint matrix.

The matrix of X is the mixed ab-representation is defined by

$$X = \sum_{a'b'} \langle a'|X|b'\rangle M(a', b') \tag{1.59}$$

where

$$\langle a'|X|b'\rangle = \operatorname{tr} XM(b', a') \quad . \tag{1.60}$$

The rule of multiplication for matrices in mixed representations is

$$\langle a'|XY|c'\rangle = \sum_{b'} \langle a'|X|b'\rangle\langle b'|Y|c'\rangle \ . \tag{1.61}$$

On placing $X = Y = 1$ we encounter the composition property of transformation functions, since

$$\langle a'|1|b'\rangle = \operatorname{tr} M(b', a') = \langle a'|b'\rangle. \tag{1.62}$$

If we set X or Y equal to 1 , we obtain examples of the connection between the matrices of a given operator in various representations. The general result can be derived from the linear relations among measurement symbols. Thus,

$$\begin{aligned} \langle a'|X|d'\rangle &= \operatorname{tr} X M(d', a') \\ &= \operatorname{tr} X \sum_{b'c'} \langle c'|d'\rangle\langle a'|b'\rangle M(c', b') \\ &= \sum_{b'c'} \langle a'|b'\rangle\langle b'|X|c'\rangle\langle c'|d'\rangle \quad . \end{aligned} \tag{1.63}$$

The adjoint of an operator X, displayed in the mixed ab-basis, appears in the ba-basis with the matrix

$$\langle b'|X^\dagger|a'\rangle = \langle a'|X|b'\rangle^* \quad . \tag{1.64}$$

1.10 VARIATIONS OF TRANSFORMATION FUNCTIONS

As an application of mixed representations, we present an operator equivalent of the fundamental properties of transformation functions:

$$\begin{gathered} \sum_{b'} \langle a'|b'\rangle\langle b'|c'\rangle = \langle a'|c'\rangle \\ \langle a'|b'\rangle^* = \langle b'|a'\rangle \quad , \end{gathered} \tag{1.65}$$

which is achieved by a differential characterization of the transformation functions. If $\delta\langle a'|b'\rangle$

and $\delta\langle b'|c'\rangle$ are any conceivable infinitesmal alteration of the corresponding transformation functions, the implied variation of $\langle a'|c'\rangle$ is

$$\delta\langle a'|c'\rangle = \sum_{b'} [\delta\langle a'|b'\rangle \langle b'|c'\rangle + \langle a'|b'\rangle \delta\langle b'|c'\rangle] \quad , \tag{1.66}$$

and also

$$\delta\langle a'|b'\rangle^* = \delta\langle b'|a'\rangle \quad . \tag{1.67}$$

One can regard the array of numbers $\delta\langle a'|b'\rangle$ as the matrix of an operator in the ab-representation. We therefore write

$$\delta\langle a'|b'\rangle = i \langle a'| \delta W_{ab} |b'\rangle \quad , \tag{1.68}$$

which is the definition of an infinitesimal operator δW_{ab}. If infinitesimal operators δW_{bc} and δW_{ac} are defined similarly, the differential property (1.66) becomes the matrix equation

$$\langle a'| \delta W_{ac} |c'\rangle = \sum_{b'} [\langle a'| \delta W_{ab} |b'\rangle\langle b'|c'\rangle + \langle a'|b'\rangle\langle b'| \delta W_{bc} |c'\rangle], \tag{1.69}$$

from which we infer the operator equation

$$\delta W_{ac} = \delta W_{ab} + \delta W_{bc} \quad . \tag{1.70}$$

Thus the multiplicative composition law of transformation functions is expressed by an additive composition law for the infinitesimal operators δW.

On identifying the a- and b- descriptions in (1.70), we learn that

$$\delta W_{aa} = 0 \qquad (1.71)$$

or

$$\delta \langle a' | a'' \rangle = 0 \quad , \qquad (1.72)$$

which expresses the fixed numerical values of the transformation function

$$\langle a' | a'' \rangle = \delta(a', a'') \quad . \qquad (1.73)$$

Indeed, the latter is not an independent condition on transformation functions but is implied by the composition property and the requirement that transformation functions, as matrices, be nonsingular. If we identify the a- and c- descriptions we are informed that

$$\delta W_{ba} = -\delta W_{ab} \quad . \qquad (1.74)$$

Now

$$\begin{aligned} \delta \langle a' | b' \rangle^* &= -i \, \langle a' | \delta W_{ab} | b' \rangle^* \\ &= -i \, \langle b' | \delta W_{ab}{}^\dagger | a' \rangle \quad , \end{aligned} \qquad (1.75)$$

which must equal

$$\delta\langle b'|a'\rangle = i\langle b'|\delta W_{ba}|a'\rangle , \tag{1.76}$$

and therefore

$$\delta W_{ab}{}^{\dagger} = -\delta W_{ba} = \delta W_{ab} . \tag{1.77}$$

The complex conjugate property of transformation functions is thus expressed by the statement that the infinitesimal operators δW are Hermitian.

1.11 EXPECTATION VALUE

The expectation value of property A for systems in the state b' is the average of the possible values of A, weighted by the probabilities of occurence that are characteristic of the state b' . On using (1.33) to write the probability formula as

$$p(a' , b') = \operatorname{tr} M(a')M(b') , \tag{1.78}$$

the expectation value becomes

$$\langle A\rangle_{b'} = \sum_{a'} a'p(a', b') = \operatorname{tr} AM(b') = \langle b'|A|b'\rangle \quad , \tag{1.79}$$

where the operator A is

$$A = \sum_{a'} a'M(a') \ . \tag{1.80}$$

The correspondence thus obtained between operators and physical quantities is such that a function $f(A)$ of the property A is assigned the operator $f(A)$, and the operators associated with a complete set of compatible physical quantities form a complete set of commuting Hermitian operators. In particular, the function of A that exhibits the value unity in the state a', and zero otherwise, is characterized by the operator $M(a')$.

1.12 ADDENDUM: NON-SELECTIVE MEASUREMENTS†

† Reproduced from the Proceedings of the National Academy of Sciences Vol. 45, pp. 1552-1553 (1959).

The physical operation symbolized by $M(a')$ involves the functioning of an apparatus capable of separating an ensemble into subensembles that are distinguished by the various values of a', together with the act of selecting one subensemble and rejecting the others. The measurement process prior to the stage of selection, which we call a nonselective measurement, will now be considered for the purpose of finding its symbolic counterpart. It is useful to recognize a general quantitative interpretation attached to the measurement symbols. Let a system in the state c' be subjected to the selective $M(b')$ measurement and then to an A-measurement. The probability that the system will exhibit the value b' and then a', for the respective properties, is given by

$$p(a',b',c') = p(a',b')p(b',c') = |\langle a'|b'\rangle\langle b'|c'\rangle|^2 \\ = |\langle a'|M(b')|c'\rangle|^2.$$

If, in contrast, the intermediate B-measurement accepts all systems without discrimination, which is equivalent to performing no B-measurement, the relevant probability is

$$p(a',1,c') = |\langle a'|c'\rangle|^2 \\ = |\langle a'|\sum_{b'} M(b')|c'\rangle|^2.$$

There are examples of the relation between the symbol of any selective measurement and a corresponding probability,

$$p(a', \ ,c') = |\langle a'|M|c'\rangle|^2.$$

Now let the intervening measurement be nonselective, which is to say that the apparatus functions but no selection of systems is performed. Accordingly,

$$p(a',b,c') = \sum_{b'} p(a',b')p(b',c') \\ = \sum_{b'} |\langle a'|M(b')|c'\rangle|^2$$

which differs from

$$p(a',1,c') = |\sum_{b'}\langle a'|M(b')|c'\rangle|^2$$

by the absence of interference terms between different b' states. This indicates that the symbol to be associated with the nonselective B-measurement is

$$M_b = \sum_{b'} e^{i\varphi_{b'}} M(b')$$

where the real phases $\varphi_{b'}$ are independent, randomly distributed quantities. The uncontrollable nature of the disturbance produced by a measurement thus finds its mathematical expression in these random phase factors. Since a nonselective measurement does not discard systems we must have

$$\sum_{a'} p(a',b,c') = 1$$

which corresponds to the unitary property of the M_b operators,

$$M_b\dagger M_b = M_b M_b\dagger = 1.$$

It should also be noted that, within this probability context, the symbols of the elementary selective measurements are derived from the nonselective symbol by replacing all but one of the phases by positive infinite imaginary numbers, which is an absorptive description of the process of rejecting subensembles.

The general probability statement for successive measurements is

$$p(a',b',\ldots s',t') = |\langle a'|M(b')\ldots M(s')|t'\rangle|^2$$

which is applicable to any type of observation by inserting the appropriate measurement symbol. Other versions are

$$p(a',\ldots t') = \langle t'|(M(a')\ldots M(s'))\dagger(M(a')\ldots M(s'))|t'\rangle$$

and

$$p(a',\ldots t') = tr(M(a')\ldots M(t'))\dagger(M(a')\ldots M(t')),$$

each of which can also be extended to all types of selective measurements, and to nonselective measurements (the adjoint form is essential here). The expectation value construction shows that a quantity which equals unity if the properties A, $B, \ldots S$ successively exhibit, in the sinistral sense, the values a', b', $\ldots s'$, and is zero otherwise, is represented by the Hermitian[3] operator $(M(a')\ldots M(s'))\dagger$-$(M(a')\ldots M(s'))$.

Measurement is a dynamical process, and yet the only time concept that has been used is the primitive relationship of order. A detailed formulation of quantum dynamics must satisfy the consistency requirement that its description of the interactions that constitute measurement reproduces the symbolic characterizations that have emerged at this elementary stage. Such considerations make explicit reference to the fact that all measurement of atomic phenomena ultimately involves the amplification of microscopic effects to the level of macroscopic observation.

Further analysis of the measurement algebra leads to a geometry associated with the states of systems.

[1] This development has been presented in numerous lecture series since 1951, but is heretofore unpublished.

[2] Here we bypass the question of the utility of the real number field. According to a comment in THESE PROCEEDINGS, **44,** 223 (1958), the appearance of complex numbers, or their real equivalents, may be an aspect of the fundamental matter-antimatter duality, which can hardly be discussed at this stage.

[3] Compare P. A. M. Dirac, *Rev. Mod. Phys*, **17,** 195 (1945), where non-Hermitian operators and complex "probabilities" are introduced.

CHAPTER TWO
THE GEOMETRY OF STATES

2.1 THE NULL STATE

The uncontrollable disturbance attendant upon a measurement implies that the act of measurement is indivisible. That is to say, any attempt to trace the history of a system during a measurement process usually changes the nature of the measurement that is being performed. Hence, to conceive of a given selective measurement $M(a', b')$ as a compound measurement is without physical implication. It is only of significance that the first stage selects systems in the state b', and that the last one produces them in the state a'; the interposed states are without meaning for the measurement as a whole. Indeed, we can even invent a non-physical state to serve as the intermediary. We shall call this mental construct the null state 0, and write

$$M(a', b') = M(a', 0)M(0, b') . \qquad (2.1)$$

The measurement process that selects a system in the state b' and produces it in the null state,

$$M(0, b') = \Phi(b') , \qquad (2.2)$$

can be described as the annihilation of a system in the state b'; and the production of a system in the state a' following its selection from the null state,

$$M(a', 0) = \Psi(a') , \qquad (2.3)$$

can be characterized as the creation of a system

in the state a'. Thus, the content of (2.1) is the indiscernability of $M(a', b')$ from the compound process of the annihilation of a system in the state b' followed by the creation of a system in the state a',

$$M(a', b') = \Psi(a')\Phi(b') . \tag{2.4}$$

The extension of the measurement algebra to include the null state supplies the properties of the Ψ and Φ symbols. Thus

$$\Psi(a')^{\dagger} = \Phi(a') , \quad \Phi(b')^{\dagger} = \Psi(b') , \tag{2.5}$$

and

$$\begin{aligned} \Psi(a')\Psi(b') &= \Phi(a')\Phi(b') = 0 , \\ M(a', b')\Phi(c') &= \Psi(a')M(b', c') = 0 , \end{aligned} \tag{2.6}$$

whereas

$$\begin{aligned} M(a', b')\Psi(c') &= \langle b'|c'\rangle\Psi(a') , \\ \Phi(a')M(b', c') &= \langle a'|b'\rangle\Phi(c') , \end{aligned} \tag{2.7}$$

and

$$\Phi(a')\Psi(b') = \langle a'|b'\rangle M(0) . \tag{2.8}$$

Some properties of $M(0)$ are

$$\Psi(a')M(0) = \Psi(a') ,$$
$$M(0)\Phi(a') = \Phi(a') , \tag{2.9}$$

and

$$M(0)\Psi(a') = \Phi(a')M(0) = 0 . \tag{2.10}$$

Furthermore, in the extended measurement algebra,

$$1 = \sum_{a'} M(a') + M(0) . \tag{2.11}$$

The fundamental arbitrariness of measurement symbols expressed by the substitution (1.34),

$$M(a' , b') \to e^{-i\varphi(a')} M(a' , b') e^{i\varphi(b')} , \tag{2.12}$$

implies the accompanying substitution

$$\Psi(a') \to e^{-i\varphi(a')} \Psi(a') ,$$
$$\Phi(b') \to e^{i\varphi(b')} \Phi(b') , \tag{2.13}$$

in which we have effectively removed $\varphi(0)$ by expressing all other phases relative to it.

2.2 RECONSTRUCTION OF THE MEASUREMENT ALGEBRA

The characteristics of the measurement operators $M(a', b')$ can now be derived from those of the Ψ and Φ symbols. Thus

$$M(a', b')^{\dagger} = \Phi(b')^{\dagger}\,\Psi(a')^{\dagger} = \Psi(b')\Phi(a') = M(b', a') , \tag{2.14}$$

and

$$\operatorname{tr} M(b',a') = \operatorname{tr}\Psi(b')\Phi(a') = \operatorname{tr}\Phi(a')\Psi(b') = \langle a'|b'\rangle \operatorname{tr} M(0) = \langle a'|b'\rangle , \tag{2.15}$$

while

$$M(a', b')M(c', d') = M(a', b')\Psi(c')\Phi(d') = \langle b'|c'\rangle\Psi(a')\Phi(d') = \langle b'|c'\rangle M(a', d') . \tag{2.16}$$

In addition, the substitution (2.13) transforms the measurement operators in accordance with (2.12).

The various equivalent statements contained in (2.6) show that the only significant products -- those not identically zero -- are of the form $\Psi\Phi$, $\Phi\Psi$ and $X\Psi$, ΦX , in addition to XY , where the latin symbols are operators, elements of the physical measurement algebra. According to the measurement operator construction (2.4), all operators are linear combinations of products $\Psi\Phi$,

$$X = \sum_{a'b'} \Psi(a')\langle a'|X|b'\rangle\Phi(b') \tag{2.17}$$

and the evaluation of the products $X\Psi$, ΦX, and XY reduces to the ones contained in (2.7),

$$\begin{aligned} \Psi(a')\Phi(b')\Psi(c') &= \Psi(a')\langle b'|c'\rangle , \\ \Phi(a')\Psi(b')\Phi(c') &= \langle a'|b'\rangle\Phi(c') . \end{aligned} \tag{2.18}$$

Hence, in any manipulation of operators leading to a product $\Phi\Psi$, the latter is effectively equal to a number,

$$\Phi(a')\Psi(b') = \langle a'|b'\rangle , \tag{2.19}$$

and in particular

$$\Phi(a')\Psi(a'') = \delta(a', a'') . \tag{2.20}$$

It should also be observed that, in any application of 1 as an operator we have, in effect,

$$1 = \sum_{a'} M(a') = \sum_{a'} \Psi(a')\Phi(a') . \tag{2.21}$$

Accordingly,

$$X = \sum_{a'b'} \Psi(a')\Phi(a')X\Psi(b')\Phi(b') \ , \tag{2.22}$$

which shows that

$$\Phi(a')X\Psi(b') = \langle a'|X|b'\rangle \ . \tag{2.23}$$

The bracket symbols

$$\langle a'| = \Phi(a') \ , \quad |b'\rangle = \Psi(b') \tag{2.24}$$

are designed to make this result an automatic consequence of the notation (Dirac). In the bracket notation various theorems, such as the law of matrix multiplication (1.61), or the general formula for change of matrix representation (1.63), appear as simple applications of the expression for the unit operator

$$1 = \sum_{a'} |a'\rangle\langle a'| \ . \tag{2.25}$$

2.3 VECTOR ALGEBRA

We have associated a Ψ and a Φ symbol with each of the N physical states of a description. Now the symbols of one description are linearly related to those of another description,

$$\Psi(b') = \sum_{a'} \Psi(a')\Phi(a')\Psi(b') = \sum_{a'} \Psi(a')\langle a'|b'\rangle \ , \tag{2.26}$$

and

$$\Phi(a') = \sum_{b'} \langle a'|b'\rangle\Phi(b') \quad , \tag{2.27}$$

which also implies the linear relation between measurement operators of various types. Arbitrary numerical multiples of Ψ or Φ symbols thus form the elements of two mutually adjoint algebras of dimensionality N , which are vector algebras since there is no significant multiplication of elements within each algebra. We are thereby presented with an N-dimensional geometry -- the geometry of states -- from which the measurement algebra can be derived, with its properties characterized in geometrical language. This geometry is metrical since the number $\Phi\Psi$ defines a scalar product. According to (2.20), the vectors $\Phi(a')$ and $\Psi(a')$ of the a-description provides an orthonormal vector basis or coordinate system, and thus the vector transformation equations (2.26) and (2.27) describe a change in coordinate system. The product of an operator with a vector expresses a mapping upon another vector in the same space,

$$X\Psi(b') = \sum_{a'} \Psi(a')\Phi(a')X\Psi(b') = \sum_{a'} \Psi(a')\langle a'|X|b'\rangle ,$$
$$\Phi(a')X = \sum_{b'} \langle a'|X|b'\rangle\Phi(b') . \tag{2.28}$$

The effect on the vectors of the a-coordinate system of the operator symbolizing property A ,

$$A = \sum a'\Psi(a')\Phi(a') \tag{2.29}$$

is given by

$$A\Psi(a') = a'\Psi(a') , \quad \Phi(a')A = \Phi(a')a' , \tag{2.30}$$

which characterizes $\Psi(a')$ and $\Phi(a')$ as the right and left eigenvectors, respectively, of the complete set of commuting operators A , with the eigenvalues a' . Associated with each vector algebra there is a dual algebra in which all numbers are replaced by their complex conjugates.

2.4 WAVE FUNCTIONS

The eigenvectors of a given description provide a basis for the representation of an arbitrary vector by N numbers. The abstract properties of vectors are realized by these sets of numbers, which are known as wave functions. We write

$$\Psi = \sum_{a'} |a'><a'|\Psi = \sum_{a'} |a'>\psi(a') \quad , \tag{2.31}$$

and

$$\Phi = \sum_{a'} \phi(a')<a'| \quad , \qquad \phi(a') = \Phi|a'> \quad . \tag{2.32}$$

If Φ and Ψ are in adjoint relation, $\Phi = \Psi^\dagger$, the corresponding wave functions are connected by

$$\phi(a') = \psi(a')^* \quad . \tag{2.33}$$

The scalar product of two vectors is

$$\Phi_1\Psi_2 = \sum_{a'} \Phi_1|a'><a'|\Psi_2 = \sum_{a'} \phi_1(a')\psi_2(a') \tag{2.34}$$

and, in particular,

$$\Psi^\dagger\Psi = \sum_{a'} \psi(a')^*\psi(a') \geq 0 \tag{2.35}$$

which characterizes the geometry of states as a unitary geometry. The operator $\Psi_1\Phi_2$ is represented by the matrix

$$<a'|\Psi_1\Phi_2|b'> = \psi_1(a')\phi_2(b') \quad , \tag{2.36}$$

and wave functions that represent $X\Psi$ and ΦX are

$$<a'|X\Psi = \sum_{b'} <a'|X|b'>\psi(b') \tag{2.37}$$

and

$$\Phi X|b'> = \sum_{a'} \phi(a')<a'|X|b'> \quad . \tag{2.38}$$

On placing $X = 1$, we obtain the relation between the wave functions of a given vector in two different representations,

$$\begin{aligned} \psi(a') &= \sum_{b'} <a'|b'>\psi(b') \\ \phi(b') &= \sum_{a'} \phi(a')<a'|b'> \quad . \end{aligned} \tag{2.39}$$

Note that the wave function representing $\Psi(b')$ in the a-description is

$$\begin{aligned} \psi_{b'}(a') &= \langle a'|\Psi(b') = \langle a'|b'\rangle \\ &= \phi_{a'}(b') \ . \end{aligned} \tag{2.40}$$

From the viewpoint of the extended measurement algebra, ϕ and ψ wave functions are matrices with but a single row, or column, respectively.

It is a convenient fiction to assert that every Hermitian operator symbolizes a physical quantity, and that every unit vector symbolizes a state. Then the expectation value of property X in the state Ψ is given by

$$\langle X\rangle_{\Psi} = \Psi^{\dagger}X\Psi = \sum_{a'a''} \psi(a')^{*}\langle a'|X|a''\rangle\psi(a'') \ . \tag{2.41}$$

In particular, the probability of observing the values a' in an A-measurement performed on systems in the state Ψ is

$$\begin{aligned} p(a', \Psi) &= \langle M(a')\rangle_{\Psi} = \Psi^{\dagger}|a'\rangle\langle a'|\Psi \\ &= |\psi(a')|^{2} \ . \end{aligned} \tag{2.42}$$

2.5 UNITARY TRANSFORMATIONS

The automophisms of the unitary geometry of states are produced by the unitary transformations

$$\bar{\Phi} = \Phi U \ , \quad \bar{\Psi} = U^{-1}\Psi \ , \quad \bar{X} = U^{-1}XU \tag{2.43}$$

applied to every vector and operator, where the unitary operator U obeys

$$U^{\dagger} = U^{-1} \ . \tag{2.44}$$

All algebraic relations and adjoint connections among vectors and operators are preserved by this transformation. Two successive unitary transformations form a unitary transformation, and the inverse of a unitary transformation is unitary - unitary transformations form a group. The application of a unitary transformation to the orthonormal basis vectors of the a-description, which are characterized by the eigenvector equation

$$\langle a'|(A - a') = 0 \tag{2.45}$$

yields orthonormal vectors

$$\langle \bar{a}'| = \langle a'|U \tag{2.46}$$

that obey the eigenvector equation

$$\langle \bar{a}'|(\bar{A} - a') = 0 \ . \tag{2.47}$$

Hence the $\langle\bar{a}'|$ are the states of a new description associated with quantities $\bar{A}$ that possess the same eigenvalue spectrum as the properties A. Since all relations among operators and vectors are preserved by the transformation, we have

$$\begin{aligned} \langle\bar{a}'|\bar{X}|\bar{a}''\rangle &= \langle a'|X|a''\rangle \\ \langle\bar{a}'|\bar{\Psi} = \langle a'|\Psi \ , &\quad \bar{\Phi}|\bar{a}'\rangle = \Phi|a'\rangle \ . \end{aligned} \tag{2.48}$$

The equivalent forms

$$\begin{aligned} \langle\bar{a}'|X|\bar{a}''\rangle &= \langle a'|UX\,U^{-1}|a''\rangle \\ \langle\bar{a}'|\Psi = \langle a'|U\Psi \ , &\quad \Phi|\bar{a}'\rangle = \Phi\,U^{-1}|a'\rangle \end{aligned} \tag{2.49}$$

exhibit the $\bar{a}$-representatives of operators and vectors as the a-representatives of associated operators and vectors.

The basis vectors of any two descriptions, with each set placed in a definite order, are connected by a unitary operator. Thus

$$\left.\begin{aligned} |a^k\rangle &= U_{ab}|b^k\rangle \\ \langle b^k| &= \langle a^k|U_{ab} \end{aligned}\right\} \quad k = 1\ldots N \ , \tag{2.50}$$

where

$$U_{ab} = \sum_{k=1}^{N} |a^k\rangle\langle b^k| \tag{2.51}$$

obeys

$$U_{ab}{}^{\dagger} = U_{ba} = U_{ab}^{-1} \quad . \tag{2.52}$$

The transformation function relating the a- and b-representations can thereby be exhibited as a matrix referring entirely to the a- or the b-representations,

$$\langle a^k|b^\ell\rangle = \langle a^k|U_{ba}|a^\ell\rangle = \langle b^k|U_{ba}|b^\ell\rangle \ , \tag{2.53}$$

and all quantities of the b-representation can be expressed as a-representatives of associated operators and vectors,

$$\begin{aligned} &\langle b^k|X|b^\ell\rangle = \langle a^k|U_{ab}XU_{ba}|a^\ell\rangle \quad . \\ &\langle b^k|\Psi = \langle a^k|U_{ab}\Psi \ , \quad \Phi|b^k\rangle = \Phi U_{ba}|a^k\rangle \ . \end{aligned} \tag{2.54}$$

If the two sets of properties A and B possess the same spectrum of values, the operators A and B are also connected by a unitary transformation. With the ordering of basis vectors established by corresponding eigenvalues we have

$$\begin{aligned} B &= \sum_k b^k |b^k\rangle\langle b^k| = \sum a^k U_{ba} |a^k\rangle\langle a^k| U_{ab} \\ &= U_{ba} A U_{ab} \quad . \end{aligned} \tag{2.55}$$

2.6 INFINITESIMAL UNITARY TRANSFORMATIONS

The definition of a unitary operator, when expressed as

$$(U-1)^\dagger (U-1) + (U-1) + (U-1)^\dagger = 0 \ , \tag{2.56}$$

shows that a unitary operator differing infinitesimally from unity has the general form

$$U = 1 + iG \ , \quad U^\dagger = U^{-1} = 1 - iG \ , \tag{2.57}$$

where G is an infinitesimal Hermitian operator. The coordinate vector transformation described by this operator is indicated by

$$\begin{aligned} \delta_a \langle a'| &= \langle \bar{a}'| - \langle a'| = \langle a'| iG \ , \\ \delta_a |a'\rangle &= |\bar{a}'\rangle - |a'\rangle = -iG|a'\rangle \ . \end{aligned} \tag{2.58}$$

Now, according to (2.49), the change of coordinate

system, in its effect upon the representatives of operators and vectors, is equivalent to a corresponding change of the operators and vectors relative to the original coordinate system. Hence

$$\delta_a \langle a'|X|a''\rangle = \langle \bar{a}'|X|\bar{a}''\rangle - \langle a'|X|a''\rangle$$
$$= \langle a'|\ \delta X\ |a''\rangle \tag{2.59}$$

and

$$\delta_a \langle a'|\Psi = \langle a'|\ \delta\Psi\ , \qquad \delta_a\ \Phi|a'\rangle = \delta\Phi\ |a'\rangle\ , \tag{2.60}$$

where

$$\delta\Psi = (U-1)\Psi = iG\Psi$$
$$\delta\Phi = \Phi(U^{-1}-1) = -\ \Phi iG\ , \tag{2.61}$$

and

$$\delta X = UX\,U^{-1} - X = \frac{1}{i}\,[X\ ,\ G]\ . \tag{2.62}$$

The rectangular bracket represents the commutator

$$[A\ ,\ B] = AB - BA\ . \tag{2.63}$$

Since all algebraic relations are preserved, the operator and vector variations are governed by rules of the type

$$\begin{aligned} \delta(XY) &= \delta X\, Y + X\, \delta Y \\ \delta(X\Psi) &= \delta X\, \Psi + X\, \delta\Psi\ . \end{aligned} \tag{2.64}$$

One must distinguish between $X + \delta X$ and

$$\bar{X} = U^{-1}XU = X - \delta X\ ; \tag{2.65}$$

the latter is the operator that exhibits the same properties relative to the $\bar{a}$-description that X possesses in the a-description. Thus the basis vectors $\langle \bar{a}'|$ are the eigenvectors of $A - \delta A$ with the eigenvalues a' .

2.7 SUCCESSIVE UNITARY TRANSFORMATIONS

In discussing successive unitary transformations, it must be recognized that a transformation which is specified by an array of numerical coefficients is symbolized by a unitary operator that depends upon the coordinate system to which it is applied. Thus, let U_1 and U_2 be the operators describing two different transformations on the same coordinate system. When the first transformation has been applied, the operator that symbolizes the second transformation, in its effect upon the coordinate system that has resulted from the initial transformation, is

$$\bar{U}_2 = U_1^{-1} U_2 U_1 \ . \tag{2.66}$$

Hence the operator that produces the complete transformation is

$$U_1 \bar{U}_2 = U_2 U_1 \ . \tag{2.67}$$

The same form with the operators of successive transformations multiplied from right to left, applies to any number of transformations. In particular, if one follows two transformations, applied in one order, by the inverse of the successive transformations in the opposite order, the unitary operator for the resulting transformation is

$$U_{[12]} = \left(U_1 U_2\right)^{-1} U_2 U_1 = U_{[21]}{}^{-1} \ . \tag{2.68}$$

When both transformations are infinitesimal,

$$U_{1,2} = 1 + iG_{1,2} \ , \tag{2.69}$$

the combined transformation described by

$$U_{[12]} = 1 + iG_{[12]} \tag{2.70}$$

is infinitesimal to the first order in each of the individual transformations,

$$\begin{aligned} G_{[12]} &= \frac{1}{i} [G_1 , G_2] \\ &= - G_{[21]} \end{aligned} \tag{2.71}$$

The infinitesimal change that the latter transformation produces in an operator is

$$\delta_{[12]}X = \delta_2\delta_1 X - \delta_1\delta_2 X , \tag{2.72}$$

which, expressed in terms of commutators, yields the operator identity (Jacobi)

$$[[X , G_1] , G_2] - [[X , G_2] , G_1] = [X , [G_1 , G_2]]. \tag{2.73}$$

2.8 UNITARY TRANSFORMATION GROUPS. TRANSLATIONS AND ROTATIONS

The continual repetition of an infinitesimal unitary transformation generates a finite unitary transformation. On writing the infinitesimal Hermitian operator G, the generator of the unitary transformation, as $\delta\tau G_{(1)}$, we find that the application of the infinitesimal transformation a number of times expressed by $\tau/\delta\tau$ yields, in the limit $\delta\tau\to 0$,

$$U(\tau) = \lim_{\delta\tau\to 0} \left(1 + i\,\delta\tau\, G_{(1)}\right)^{\tau/\delta\tau} = e^{i\tau G_{(1)}} \quad . \tag{2.74}$$

These operators form a one-parameter continuous group of unitary transformations,

$$\begin{aligned} &U(\tau_1)U(\tau_2) = U(\tau_1+\tau_2) \ , \\ &U(\tau)^{-1} = U(-\tau) \ , \quad U(0) = 1 \ . \end{aligned} \tag{2.75}$$

A number of finite Hermitian operators $G_{(1)}, \ldots, G_{(k)}$, generates a k-parameter continuous group of unitary transformations if they form a linear basis for an operator ring that is closed under the unitary transformations of the group. This requires that all commutators $[G_{(i)}, G_{(j)}]$ be linear combinations of the generating operators.

There is a fundamental continuous group of unitary transformations based upon the significance of measurements as physical operations in three-dimensional space. A measurement apparatus defines a spatial coordinate system with respect to which physical properties are specified. We express the uniformity of space by asserting that two coordinate systems, differing only in location and orientation, are intrinsically equivalent. In particular, physical quantities that are analogously defined with respect to different coordinate systems exhibit the same spectra of possible values, and the associated operators must be related by a unitary transformation. Since the totality of translations and rotations of a coordinate system form a six-parameter continuous group, we infer the existence of an isomorphic group of unitary operators.

An infinitesimal change of coordinate system is specified by stating that a point with coordinate vector $\mathbf{x}$ in the initial system is assigned the coordinate vector $\mathbf{x} - \delta\mathbf{x}$ in the new system, where

$$\delta\mathbf{x} = \delta\boldsymbol{\varepsilon} + \delta\boldsymbol{\omega} \times \mathbf{x} . \qquad (2.76)$$

The infinitesimal generator of the corresponding

unitary transformation is written

$$
\begin{aligned}
G_{\mathbf{x}} &= \delta\boldsymbol{\varepsilon} \cdot \mathbf{P} + \delta\boldsymbol{\omega} \cdot \mathbf{J} \\
&= \sum_k \delta\varepsilon_k \, P_k + \tfrac{1}{2} \sum_{k\ell} \delta\omega_{k\ell} \, J_{k\ell} \ ,
\end{aligned}
\tag{2.77}
$$

with the usual association of axial vectors and antisymmetrical tensors characteristic of three dimensions. On comparing the two ways in which a pair of infinitesimal coordinate changes can be performed, in the manner of (2.68), we find that

$$
\begin{aligned}
\delta_{[12]}\boldsymbol{\varepsilon} &= \delta\boldsymbol{\omega}_1 \times \delta\boldsymbol{\varepsilon}_2 - \delta\boldsymbol{\omega}_2 \times \delta\boldsymbol{\varepsilon}_1 \\
\delta_{[12]}\boldsymbol{\omega} &= \delta\boldsymbol{\omega}_1 \times \delta\boldsymbol{\omega}_2 \ ,
\end{aligned}
\tag{2.78}
$$

which requires that the associated infinitesimal generators obey the commutation relation (2.71),

$$
\begin{aligned}
&\frac{1}{i} [\delta\boldsymbol{\varepsilon}_1 \cdot \mathbf{P} + \delta\boldsymbol{\omega}_1 \cdot \mathbf{J} \ , \ \delta\boldsymbol{\varepsilon}_2 \cdot \mathbf{P} + \delta\boldsymbol{\omega}_2 \cdot \mathbf{J}] \\
&= (\delta\boldsymbol{\omega}_1 \times \delta\boldsymbol{\varepsilon}_2 - \delta\boldsymbol{\omega}_2 \times \delta\boldsymbol{\varepsilon}_1) \cdot \mathbf{P} + \delta\boldsymbol{\omega}_1 \times \delta\boldsymbol{\omega}_2 \cdot \mathbf{J} \ .
\end{aligned}
\tag{2.79}
$$

Hence,

$$\frac{1}{i}\,[P_k\ ,\ P_\ell] = 0\ ,$$

$$\frac{1}{i}\,[P_k\ ,\ J_{\ell m}] = \delta_{km}\,P_\ell - \delta_{k\ell}\,P_m\ ,$$

$$\frac{1}{i}\,[J_{k\ell}\ ,\ J_{mn}] = \delta_{\ell n}\,J_{km} - \delta_{kn}\,J_{\ell m} + \delta_{km}\,J_{\ell n} - \delta_{\ell m}\,J_{kn}\ , \tag{2.80}$$

where the last statement appears in three-dimensional vector notation as

$$\mathbf{J} \times \mathbf{J} = i\mathbf{J}\ . \tag{2.81}$$

The six Hermitian operators comprised in $\mathbf{P}$ and $\mathbf{J}$, the generators of infinitesimal translations and rotations, respectively, are identified as the operators of total linear momentum and total angular momentum. These physical quantities appear measured in certain natural units - pure numbers for angular momentum, and inverse length for linear momentum. The connection between such atomic units and the conventional macroscopic standards must be found empirically. If the latter are to be employed, a conversion factor should be introduced, which involves the replacement of $\mathbf{P}$ and $\mathbf{J}$ with $\hbar^{-1}\mathbf{P}$ and $\hbar^{-1}\mathbf{J}$, respectively. The constant $\hbar$ possesses the dimensions of action, and its measured value is

$$\hbar = 1.0545 \times 10^{-27} \text{ erg sec.} \tag{2.82}$$

The natural units are preferable for general theoretical investigations and will be used here.

2.9 REFLECTIONS

The continuous group of transformations among kinematically equivalent coordinate systems can be enlarged by the operation of reflecting the positive sense of every spatial coordinate axis. We associate with this change of description the unitary reflection operator R,

$$R^{\dagger}R = 1 . \tag{2.83}$$

A reflection, followed by the infinitesimal displacement $\delta\boldsymbol{\varepsilon}$, $\delta\boldsymbol{\omega}$ is equivalent to first performing the displacement $-\delta\boldsymbol{\varepsilon}$, $\delta\boldsymbol{\omega}$, and then the reflection. Accordingly,

$$\begin{aligned}[1 + i(\delta\boldsymbol{\varepsilon} \cdot \mathbf{P} + \delta\boldsymbol{\omega} \cdot \mathbf{J})]R \\ = R[1 + i(-\delta\boldsymbol{\varepsilon} \cdot \mathbf{P} + \delta\boldsymbol{\omega} \cdot \mathbf{J})] ,\end{aligned} \tag{2.84}$$

or

$$R^{-1}P_kR = -P_k , \quad R^{-1}J_{k\ell}R = J_{k\ell} . \tag{2.85}$$

2.10 CONTINUOUS SPECTRA

The general mathematical structure of quantum mechanics as the symbolic expression of the laws of atomic measurement has been developed in the context of physical systems possessing a finite number of states. We shall comment only briefly on the extension of these considerations to systems with infinite numbers of states, and properties exhibiting a continuous spectrum of possible values. In any measurement of such a property, systems displaying values within a certain range are selected, and the concept of state now refers to the specification of a complete set of compatible quantities within arbitrarily small intervals about prescribed values. We symbolize such states by $|a'\rangle_\Delta$, $_\Delta\langle a'|$, and express their completeness by

$$1 = \sum_{a'} |a'\rangle_\Delta \, _\Delta\langle a'| . \tag{2.86}$$

The change of normalization,

$$|a'>_{\Delta} = (\Delta a')^{\frac{1}{2}} \, |a'>$$
$$_{\Delta}<a'| = (\Delta a')^{\frac{1}{2}} \, <a'| \, , \tag{2.87}$$

in which $\Delta a'$ is the product of eigenvalue intervals for each continuous property, now yields, in the limit $\Delta a' \to 0$,

$$1 = \int |a'> da' <a'| \, , \tag{2.88}$$

if all members of set A have continuous spectra. We deduce, for any vector Ψ represented by an arbitrary wave function

$$\psi(a') = <a'|\Psi \, , \tag{2.89}$$

that

$$\psi(a') = \int <a'|a''> da'' \, \psi(a'') \, , \tag{2.90}$$

which is the operational definition of the delta function ,

$$<a'|a''> = \delta(a'-a'') \, . \tag{2.91}$$

This is the continuum analogue of the property

$$\psi(a') = \sum_{a''} \delta(a' , a'') \psi(a'') \tag{2.92}$$

and generally, in all formal relations referring to continuous spectra, integrals replace summations. In particular, the probability that a measurement on a system in the state Ψ will yield one of a set of states appears as an integral over that set, $\int da' \, |\psi(a')|^2$, so that

$$dp(a' , \Psi) = da' \, |\psi(a')|^2 \tag{2.93}$$

can be described as the probability of encountering a system with properties A in the infinitesimal range da' about a' .

2.11 ADDENDUM: OPERATOR SPACE†

---

† Reproduced from the Proceedings of the National Academy of Sciences, Vol. 46, pp 261-265 (1960).

The geometry of states provides the elements of the measurement algebra with the geometrical interpretation of operators on a vector space. But operators considered in themselves also form a vector space, for the totality of operators is closed under addition and under multiplication by numbers. The dimensionality of this operator space is N^2 according to the number of linearly independent measurement symbols of any given type. A unitary scalar product is defined in the operator space by the number

$$<X|Y> = tr(X^\dagger Y) = <Y^\dagger|X^\dagger>,$$

which has the properties

$$\begin{aligned} <X|Y>^* &= <Y|X> \\ <X|X> &\geq 0. \end{aligned}$$

The trace evaluation

$$\text{tr}\, M(b', a')M(a'', b'') = \delta(a', a'')\delta(b', b'')$$

characterizes the $M(a', b')$ basis as orthonormal.

$$<M(a', b')|M(a'', b'')> = \delta(a'b', a''b''),$$

and the general linear relation between measurement symbols,

$$M(c', d') = \sum_{a'b'} <a'|c'><d'|b'>\, M(a', b'),$$

can now be viewed as the transformation connecting two orthonormal bases. This change of basis is described by the transformation function

$$<a'b'|c'd'> \equiv <M(a'b')|M(c', d') = <a'|c'>\,<d'|b'>,$$

which is such that

$$<a'b'|c'd'> = <d'c'|b'a'>.$$

and

$$\begin{aligned} <a'b'|c'd'>^* &= <c'd'|a'b'> \\ &= <b'a'|d'c'>. \end{aligned}$$

One can also verify the composition property of transformation functions,

$$\sum_{c'd'} <a'b'|c'd'>\,<c'd'|e'f'> = <a'b'|e'f'>.$$

The probability relating two states appears as a particular type of operator space transformation function,

$$\begin{aligned} p(a', b') &= <a'|b'>\,<b'|a'> \\ &= <a'a'|b'b'>. \end{aligned}$$

Let $X(\alpha)$, $\alpha = 1\,..\,N^2$, be the elements of an arbitrary orthonormal basis,

$$<X(\alpha)|X(\alpha')> = \delta(\alpha, \alpha').$$

The connection with the $M(a', b')$ basis is described by the transformation function

$$\begin{aligned} <a'b'|\alpha> &= \text{tr}\, M(b', a')X(\alpha) \\ &= <a'|X(\alpha)|b'>. \end{aligned}$$

We also have

$$<\alpha|a'b'> = <a'b'|\alpha>^* \\ = <b'|X(\alpha)\dagger|a'>$$

and the transformation function property

$$\sum_\alpha <a'b'|\alpha> <\alpha|a''b''> = \delta(a'b', a''b'')$$

acquires the matrix form

$$\sum_\alpha <a'|X(\alpha)|b'> <b''|X(\alpha)^\dagger|a''> = \delta(a', a'')\delta(b', b'').$$

If we multiply the latter by the b-matrix of an arbitrary operator Y, the summation with respect to b' and b'' yields the a-matrix representation of the operator equation

$$\sum_\alpha X(\alpha)YX(\alpha)^\dagger = 1 \text{ tr } Y,$$

the validity of which for arbitrary Y is equivalent to the completeness of the operator basis $X(\alpha)$. Since the operator set $X(\alpha)\dagger$ also forms an orthonormal basis we must have

$$\sum_\alpha X(\alpha)\dagger YX(\alpha) = 1 \text{ tr } Y,$$

and the particular choice $Y = 1/N$ gives

$$\frac{1}{N}\sum_\alpha X(\alpha)X(\alpha)\dagger = \frac{1}{N}\sum_\alpha X(\alpha)\dagger\, X(\alpha) = 1.$$

The expression of an arbitrary operator relative to the orthonormal basis $X(\alpha)$,

$$X = \sum_\alpha X(\alpha)x(\alpha),$$

defines the components

$$x(\alpha) = <X(\alpha)|X> \equiv <\alpha|X>$$

For the basis $M(a', b')$, the components are

$$x(a'b') = tr\, M(b', a')X \\ = <a'|X|b'>,$$

the elements of the ab-matrix representation of X. The scalar product in operator space is evaluated as

$$<X|Y> = \sum_\alpha x(\alpha)^*y(\alpha)$$

and

$$<X|X> = \sum_\alpha |x(\alpha)|^2 \geq 0.$$

On altering the basis the components of a given operator change in accordance with

$$x(\alpha) = \sum_\beta <\alpha|\beta> x(\beta).$$

For measurement symbol bases this becomes the law of matrix transformation.

There are two aspects of the operator space that have no counterpart in the state spaces—the adjoint operation and the multiplication of elements are defined in the same space. Thus

$$X(\alpha) = \sum_{\beta} (\alpha\beta)X(\beta)\dagger$$
$$(\alpha\beta) = (\beta\alpha) = \text{tr } X(\alpha)\, X(\beta),$$

and

$$X(\alpha)X(\beta) = \sum_{\gamma} (\alpha\beta\gamma)X(\gamma)\dagger$$
$$= \sum_{\gamma} X(\gamma) < \gamma|\alpha\beta >$$

where

$$(\alpha\beta\gamma) = (\beta\gamma\alpha) = (\gamma\alpha\beta)$$
$$= \text{tr } X(\alpha)X(\beta)X(\gamma)$$

and

$$<\gamma|\alpha\beta> = \text{tr } X(\gamma)\dagger X(\alpha)X(\beta).$$

Some consequences are

$$<\alpha|X>^* = \sum_{\beta} (\alpha\beta) < \beta|X\dagger>,$$
$$<\gamma|XY> = \sum_{\alpha\beta} <\gamma|\alpha\beta> <\alpha|X> <\beta|Y>,$$

which generalize the adjoint and multiplication properties of matrices. The elements of the operator space appear in the dual role of operator and operand on defining matrices by

$$<\alpha|X|\alpha''> = <\alpha|XX(\alpha'')>$$
$$= \sum_{\alpha'} <\alpha|\alpha'\alpha''> <\alpha'|X>.$$

The measurement symbol bases are distinguished in this context by the complete reducibility of such matrices, in the sense of

$$<a'b'|X|a''b''> = <a'|X|a''> \delta(b', b'').$$

Otherwise expressed, the set of N measurement symbols $M(a', b')$, for fixed b', or fixed a', are left and right ideals, respectively, of the operator space.

The possibility of introducing Hermitian orthonormal operator bases is illustrated by the set

$$a' \neq a'': \quad \begin{array}{l} 2^{-1/2}[M(a', a'') + M(a'', a')] \\ 2^{-1/2}\, i[M(a', a'') - M(a'', a')]\,, \; M(a'). \end{array}$$

For any such basis

$$<\alpha|\alpha'> = (\alpha\alpha') = \delta(\alpha, \alpha')$$

and

$$<\alpha|X>^* = <\alpha|X\dagger>,$$

which implies that a Hermitian operator X has real components relative to a Hermitian basis, and therefore

$$<X|X> = \sum_{\alpha} x(\alpha)^2 \geq 0.$$

Thus the subspace of Hermitian operators is governed by Euclidean geometry, and a change of basis is a real orthogonal transformation,

$$X(\alpha) = \sum_{\beta} (\alpha\beta)X'(\beta).$$

When the unit operator (multiplied by $N^{-1/2}$) is chosen as a member of such bases it defines an invariant subspace, and the freedom of orthogonal transformation refers to the $N^2 - 1$ basis operators of zero trace.

Important examples of orthonormal operator bases are obtained through the study of unitary operators.

[1] Schwinger, J., these PROCEEDINGS, **45**, 1542 (1959).

2.12 ADDENDUM: UNITARY OPERATOR BASES†

† Reproduced from the Proceedings of the National Academy of Sciences, Vol. 46, pp. 570-579 (1960).

Reprinted from the Proceedings of the NATIONAL ACADEMY OF SCIENCES
Vol. 46, No. 4, pp. 570–579. April, 1960.

*UNITARY OPERATOR BASES**

BY JULIAN SCHWINGER

HARVARD UNIVERSITY

Communicated February 2, 1960

To qualify as the fundamental quantum variables of a physical system, a set of operators must suffice to construct all possible quantities of that system. Such operators will therefore be identified as the generators of a complete operator basis. Unitary operator bases are the principal subject of this note.[1]

Two state vector space coordinate systems and a rule of correspondence define a unitary operator. Thus, given the two ordered sets of vectors $\langle a^k|$, $\langle b^k|$, $k = 1..N$, and their adjoints, we construct

$$U_{ab} = \sum_{k=1}^{N} |a^k\rangle \langle b^k|$$

$$U_{ba} = \sum_{k=1}^{N} |b^k\rangle \langle a^k|$$

which are such that

$$\langle a^k| U_{ab} = \langle b^k|, \quad U_{ab}|b^k\rangle = |a^k\rangle$$

$$\langle b^k| U_{ba} = \langle a^k|, \quad U_{ba}|a^k\rangle = |b^k\rangle$$

and

$$U_{ab}U_{ba} = U_{ba}U_{ab} = 1, \quad U_{ab}\dagger = U_{ba},$$

implying the unitary property

$$U\dagger = U^{-1}$$

for both U_{ab} and U_{ba}. If a third ordered coordinate system is given, $\langle c^k|$, $k = 1..N$, we can similarly define the unitary operators U_{ac}, U_{bc}, which obey the composition property

$$U_{ab}U_{bc} = U_{ac}.$$

A unitary operator is also implied by two orthonormal operator bases in a given space, that have the same multiplication properties:

$$X(\alpha')X(\alpha'') = \sum_{\alpha=1}^{N^2} X(\alpha)\,\langle\alpha|\,\alpha'\alpha''\rangle$$
$$Y(\alpha')Y(\alpha'') = \sum_{\alpha} Y(\alpha)\,\langle\alpha|\,\alpha'\alpha''\rangle.$$

Let us define

$$U = \lambda \sum_{\alpha} X(\alpha)Y(\alpha)\dagger$$

and observe that

$$X(\alpha')U = \lambda \sum_{\alpha\alpha''} X(\alpha)\,\langle\alpha|\,\alpha'\alpha''\rangle\, Y(\alpha'')\dagger = UY(\alpha'),$$

where the latter statement follows from the remark that the $Y(\alpha)\dagger$ form an orthonormal basis and therefore

$$Y(\alpha)\dagger Y(\alpha') = \sum_{\alpha''} Y(\alpha'')\dagger tr Y(\alpha)\dagger Y(\alpha')Y(\alpha'')$$
$$= \sum_{\alpha''} \langle\alpha|\,\alpha'\alpha''\rangle\, Y(\alpha'')\dagger.$$

We also have the adjoint relation

$$Y(\alpha)\dagger U\dagger = U\dagger X(\alpha)\dagger$$

and in consequence

$$UU\dagger = \lambda\sum X(\alpha)Y(\alpha)\dagger U\dagger$$
$$= \lambda\sum X(\alpha)U\dagger X(\alpha)\dagger = 1\lambda tr U\dagger,$$

according to the completeness of the $X(\alpha)$ basis. Thus the operator U is unitary if we choose

$$\lambda^{-1} = tr U\dagger = \lambda^* \sum_{\alpha} \langle X(\alpha)|\, Y(\alpha)\rangle,$$

and, to within the arbitrariness of a phase constant,

$$\lambda = [\sum \langle X(\alpha)|\, Y(\alpha)\rangle]^{-1/2}$$

The converse theorems should be noted. For any unitary operator U, the orthonormal basis

$$Y\,(\alpha) = U^{-1}\,X\,(\alpha)\,U$$

obeys the same multiplication law as the $X(\alpha)$, and the $Y(\alpha)\dagger$ are given by the same linear combination of the $Y(\alpha)$ as are the $X(\alpha)\dagger$ of the $X(\alpha)$ set. In particular, if $X(\alpha)$ is a Hermitian basis, so also is $Y(\alpha)$.

We cannot refrain from illustrating these remarks for the simplest of the N^2-dimensional operator spaces, the quaternion space associated with a physical system possessing only two states. If a particular choice of these is arbitrarily designated as + and −, we obtain the four measurement symbols $M(\pm, \pm)$, and can then introduce a Hermitian orthonormal operator basis

$$X(\alpha) = 2^{-1/2}\sigma_\alpha \quad \alpha = 0,..3$$

such that

$$\sigma_0 = 1.$$

Accordingly, the three operators σ_k, $k = 1, 2, 3$, obey

$$tr\ \sigma_k = 0, \quad {}^1/_2 tr\ \sigma_k\sigma_l = \delta_{kl},$$

and an explicit construction is given by

$$\sigma_1 = M(+, -) + M(-, +),\ \sigma_2 = -iM(+, -) + iM(-, +),$$
$$\sigma_3 = M(+, +) - M(-, -),$$

the coefficients of which constitute the well-known Pauli matrices. With these definitions, the multiplication properties of the σ operators can be expressed as

$$\sigma_k\sigma_l = \delta_{kl} + i\sum_m \epsilon_{klm}\sigma_m,$$

or, equivalently, by the additional trace

$${}^1/_2 tr\,\sigma_k\sigma_l\sigma_m = i\epsilon_{klm}$$

where ϵ_{klm} is the alternating symbol specified by $\epsilon_{123} = +1$. If we now introduce any other Hermitian orthonormal operator basis $Y(\alpha) = 2^{-1/2}\bar{\sigma}_\alpha$, with $\bar{\sigma}_0 = 1$, the resulting three-dimensional orthonormal basis transformation

$$\bar{\sigma}_k = \sum_{l=1}^{3} r_{kl}\sigma_l, \qquad k = 1, 2, 3$$

is real and orthogonal,

$$r^* = r,\ r^T r = 1.$$

The multiplication properties of the σ-basis assert that

$${}^1/_2\ tr\ \bar{\sigma}_k\bar{\sigma}_l\bar{\sigma}_m = i\epsilon_{klm}\ det\ r,$$

where, characteristic of an orthogonal transformation, $det\ r = \pm 1$. If the orthogonal transformation is proper, the multiplication properties of the $\bar{\sigma}$-basis coincide with those of the σ-basis, while with an improper transformation the opposite sign of i is effectively employed in evaluating the $\sigma_k\bar{\sigma}_l$ products. Hence only in the first situation, that of a pure rotation, does a unitary operator exist such that

$$\sum_l r_{kl}\sigma_l = U^{-1}\sigma_k U.$$

The unitary operator is constructed explicitly as[2]

$$U = \lambda^1/_2 \sum_{\alpha=0}^{3} \sigma_\alpha\bar{\sigma}_\alpha$$
$$= \lambda^1/_2[1 + tr\ r + i\sum_{m=1}^{3} r_{kl}\epsilon_{klm}\sigma_m]$$

with

$$\lambda = (1 + tr\ r)^{-1/2}.$$

Let us return to the definition of a unitary operator through the mapping of one coordinate system or another, and remark that the two sets of vectors can be identical, apart from their ordering. Thus, consider the definition of a unitary operator V by the cyclic permutation

$$\langle a^k | V = \langle a^{k+1} |, \, k = 1..N$$

where

$$\langle a^{N+1} | = \langle a^1 |,$$

which indicates the utility of designating the same state by any of the integers that are congruent with respect to the modulus N. The repetition of V defines linearly independent unitary operators,

$$\langle a^k | V^n = \langle a^{k+n} |,$$

until we arrive at

$$\langle a^k | V^N = \langle a^{k+N} | = \langle a^k |.$$

Thus

$$V^N = 1$$

is the minimum equation, the polynomial equation of least degree obeyed by this operator, which we characterize as being of period N. The eigenvalues of V obey the same equation and are given by the N distinct complex numbers

$$v' = e^{\frac{2\pi i k}{N}} = v^k, \qquad k = 0, ..N - 1.$$

Unitary operators can be regarded as complex functions of Hermitian operators, and the entire spectral theory of Hermitian operators can be transferred to them. If the unitary operator V has N distinct eigenvalues, its eigenvectors constitute an orthonormal coordinate system. The adjoint of a right eigenvector $|v'\rangle$ is the left eigenvector $\langle v'|$ associated with the same eigenvalue, and the products

$$|v'\rangle \langle v'| = M(v')$$

have all the properties required of measurement symbols. Now let us observe that the factorization of the minimum equation for V that is given by

$$[(V/v') - 1] \sum_{l=0}^{N-1} (V/v')^l = 0$$

permits the identification of the Hermitian operator

$$M(v') = 1/N \sum (v')^{-l} V^l$$

to within the choice of the factor N^{-1}, which is such that

$$\langle v' | M(v') = \langle v' |.$$

On multiplying $M(v^k)$ by $\langle a^N|$ and using the defining property of V, we obtain

$$\langle a^N | v^k \rangle \langle v^k | = 1/N \sum_l \langle a^l | e^{-\frac{2\pi i}{N} kl}$$

from which follows

$$|\langle a^N|v^k\rangle|^2 = 1/N.$$

Then, with a convenient phase convention for $\langle a^N|v^k\rangle$ we get

$$\langle v^k| = N^{-1/2}\sum_l \langle a^l|e^{-\frac{2\pi i}{N}kl}$$

which is also expressed by

$$\langle v^k|a^l\rangle = N^{-1/2}e^{-\frac{2\pi i}{N}kl},\ \langle a^l|v^k\rangle = N^{-1/2}e^{\frac{2\pi i}{N}kl},$$

the elements of the transformation functions that connect the given coordinate system with the one supplied by the eigenvectors of the unitary operator that cyclically permutes the vectors of the given system.

Turning to the new coordinate system $\langle v^k|$, we define another unitary operator by the cyclic permutation of this set. It is convenient to introduce U such that

$$\langle v^k|U^{-1} = \langle v^{k+1}|$$

which is equivalent to

$$\langle v^k|U = \langle v^{k-1}|.$$

This operator is also of period N,

$$U^N = 1,$$

and has the same spectrum as V,

$$u' = e^{\frac{2\pi ik}{N}} = u^k,\ k = 0..N-1.$$

After using the property $U^{N-l} = U^{-l}$ to write the corresponding measurement symbol as

$$M(u') = 1/N\sum_{l=0}^{N-1}(u')^l U^{-l},$$

we follow the previous procedure to construct the eigenvectors of U,

$$\langle u^k| = N^{-1/2}\sum_l \langle v^l|e^{\frac{2\pi i}{N}kl} = \sum_l \langle a^k|v^l\rangle\langle v^l| = \langle a^k|.$$

Thus, the original coordinate system is regained and our results can now be stated as the reciprocal definition of two unitary operators and their eigenvectors,

$$\langle u^k|V = \langle u^{k+1}|$$

$$\langle v^k|U^{-1} = \langle v^{k+1}|.$$

The relation between the two coordinate systems is given by

$$\langle u^k|v^l\rangle = N^{-1/2}e^{\frac{2\pi i}{N}kl},\ \langle v^l|u^k\rangle = N^{-1/2}e^{-\frac{2\pi i}{N}kl},$$

and, supplementing, the periodic properties

$$U^N = V^N = 1,$$

we infer from the comparison

$$\langle u^k | UV = \langle u^{k+1} | u^k, \ \langle u^k | VU = \langle u^{k+1} | u^{k+1} = \langle u^{k+1} | u^k e^{\frac{2\pi i}{N}}$$

that

$$VU = e^{\frac{2\pi i}{N}} UV.$$

As a consequence of the latter result, we also have

$$V^l U^k = e^{\frac{2\pi i}{N} kl} U^k V^l.$$

Each of the unitary operators U and V is a function of a Hermitian operator that in itself forms a complete set of physical properties. It is natural to transfer this identification directly to the unitary operators which are more accessible than the implicit Hermitian operators. Accordingly, we now speak of the statistical relation between the properties U and V, as described by the probability

$$p(u', v'') = |\langle u' | v'' \rangle|^2 = 1/N.$$

The significance of this result can be emphasized by considering a measurement sequence that includes a nonselective measurement, as in

$$p(u', v, u'') = \sum_{v'} p(u', v')p(v', u'') = 1/N,$$

for this asserts that the intervening non-selective v-measurement has destroyed all prior knowledge concerning u-states. Thus the properties U and V exhibit the maximum degree of incompatibility. We shall also show that U and V are the generators of a complete orthonormal operator basis, such as the set of N^2 operators

$$X(mn) = N^{-1/2} U^m V^n, \ m, n = 0..N - 1,$$

and therefore together supply the foundation for a full description of a physical system possessing N states. Both of these aspects are implied in speaking of U and V as a *complementary* pair of operators.[3] Incidentally, there is complete symmetry between U and V, as expressed by the invariance of all properties under the substitution

$$U \rightarrow V, \ V \rightarrow U^{-1}.$$

The latter could be emphasized by choosing the elements of the operator basis as

$$N^{-1/2} e^{\frac{\pi i}{N} mn} U^m V^n = N^{-1/2} e^{-\frac{\pi i}{N} mn} V^n U^m,$$

which are invariant under this substitution when combined with $m \rightarrow n, n \rightarrow -m$.

One proof of completeness for the operator basis generated by U and V depends upon the following lemma: If an operator commutes with both U and V it is necessarily a multiple of the unit operator. Since U is complete in itself, such an operator must be a function of U. Then, according to the hypothesis of commutativity with V, for each k we have

$$0 = \langle u^k | [f(U)V - Vf(U)] | u^{k+1} \rangle = f(u^k) - f(u^{k+1}),$$

and this function of U assumes the same value for every state, which identifies it with that multiple of the unit operator. Now consider, for arbitrary Y,

$$\sum_{mn} X(mn)YX(mn)\dagger = 1/N \sum_{mn} U^m V^n Y V^{-n} U^{-m} = 1/N \sum_{mn} V^n U^m Y U^{-m} V^{-n},$$

and observe that left and right hand multiplication with U and U^{-1}, respectively, or with V and V^{-1}, only produces a rearrangement of the summations. Accordingly, this operator commutes with U and V. On taking the trace of the resulting equation, the multiple of unity is identified with $tr\ Y$, and we have obtained

$$\sum_{m,\,n\,=\,0}^{N-1} X(mn)YX(mn)\dagger = 1\ tr\ Y,$$

the statement of completeness for the N^2-dimensional operator basis $X(mn)$. Alternatively, we demonstrate that these N^2 operators are orthonormal by evaluating

$$\langle X(mn)|X(m'n')\rangle = 1/N\ tr\ U^{m'-m}V^{n'-n}$$
$$= \delta(m, m')\delta(n,n'),\ m, n, m', n' = 0..N-1.$$

The unit value for $m = m'$, $n = n'$ is evident. If $m \neq m'$, the difference $m' - m$ can assume any value between $N - 1$ and $-(N - 1)$, other than zero. When the trace is computed in the v-representation, the operator $U^{m'-m}$ changes each vector $\langle v^k|$ into the orthogonal vector $\langle v^{k+m-m'}|$ and the trace vanishes. Similarly, if $n \neq n'$ and the trace is computed in the u-representation, each vector $\langle u^k|$ is converted by $V^{n'-n}$ into the orthogonal vector $\langle u^{k+n'-n}|$ and the trace equals zero.

One application of the operator completeness property is worthy of attention. We first observe that

$$U^m V^n U V^{-n} U^{-m} = e^{\frac{2\pi i}{N}n}\, U = u^n U$$
$$U^m V^n V V^{-n} U^{-m} = e^{-\frac{2\pi i}{N}m}\, V = v^{-m} V,$$

which exhibits the unitary transformations that produce only cyclic spectral translations. Now, if Y is given as an arbitrary function of U and V, the completeness expression of the operator basis reads

$$1/N^2 \sum_{u'v'} F(u'U, v'V) = 1/N\ tr\ F$$

which is a kind of ergodic theorem, for it equates an average over all spectral translations to an average over all states. The explicit reference to operators can be removed if $F(U, V)$ is constructed from terms which, like the individual operators $X(mn)$, are ordered with U standing everywhere to the left of V. Then we can evaluate a matrix element of the operator equation, corresponding to the states $\langle u^0|$ and $|v^0\rangle$, which gives the numerical relation

$$tr\ F(U,V) = 1/N \sum_{u'v'} F(u', v').$$

It is interesting to notice that a number of the powers U^k, $k = 1 .. N - 1$, can have the period N. This will occur whenever the integers k and N have no common factor and thus the multiplicity of such operators equals $\phi(N)$, the number of integers less than and relatively prime to N. Furthermore, to every such power of U there can be associated a power V^l, also of period N, that obeys with U^k the same operator equation satisfied by U and V,

$$V^l U^k = e^{\frac{2\pi i}{N}} U^k V^l.$$

This requires the relation

$$kl = 1 \ (mod\ N),$$

and the unique solution provided by the Fermat-Euler theorem is

$$l = k^{\phi(N) - 1} \ (mod\ N).$$

The pair of operators U^k, V^l also generate the operator basis $X(mn)$, in some permuted order.

We shall now proceed to replace the single pair of complementary operators U, V by several such pairs, the individual members of which have smaller periods than the arbitrary integer N. This leads to a classification of quantum degrees of freedom in relation to the various irreducible, prime periods. Let

$$N = N_1 N_2$$

where the integers N_1 and N_2 are relatively prime, and define

$$U_1 = U^{N_2} \qquad\qquad U_2 = U^{N_1}$$

$$V_1 = V^{l_1 N_2} \qquad\qquad V_2 = V^{l_2 N_1}$$

with

$$l_1 = N_2^{\phi(N_1) - 1} \ (mod\ N_1),\ l_2 = N_1^{\phi(N_2) - 1} \ (mod\ N_2).$$

It is seen that U_1, V_1 are of period N_1, while U_2, V_2 have the period N_2, and that the two pairs of operators are mutually commutative, as illustrated by

$$V_1 U_2 = e^{\frac{2\pi i}{N} l_1 N_1 N_2} U_2 V_1 = U_2 V_1.$$

Furthermore,

$$V_1 U_1 = e^{\frac{2\pi i}{N_1}} U_1 V_1$$

$$V_2 U_2 = e^{\frac{2\pi i}{N_2}} U_2 V_2,$$

so that U_1, V_1 and U_2, V_2 constitute two independent pairs of complementary operators associated with the respective periods N_1 and N_2. We also observe that the $N = N_1 N_2$ independent powers of U can be obtained as

$$U_1^{m_1} U_2^{m_2} = U^{(m_1 N_1 + m_2 N_2)} \qquad m_1 = 0..N_1 - 1$$

$$m_2 = 0..N_2 - 1$$

since all of these are distinct powers owing to the relatively prime nature of N_1 and N_2. With a similar treatment for V, we recognize that the members of the orthonormal operator basis are given in some order by

$$X(m_1m_2n_1n_2) = N^{-1/2}U_1^{m_1}\, U_2^{m_2}\, V_1^{n_1}\, V_2^{n_2} = \prod_{j=1}^{2} X(m_jn_j),$$

where

$$X(m_jn_j) = N_j^{-1/2}U_j^{m_j}\, V_j^{n_j},\ m_j,\ n_j = 0..N_j - 1.$$

Another approach to this commutative factorization of the operator basis proceeds through the construction of the eigenvalue index $k = 0..N - 1$ with the aid of a pair of integers,

$$k = k_1N_2 + k_2 \qquad k_1 = 0..N_1 - 1$$
$$k_2 = 0..N_2 - 1.$$

Equivalently, we replace

$$u^k = v^k = e^{\frac{2\pi ik}{N}}$$

by

$$u_1^{k_1} = v_1^{k_1} = e^{\frac{2\pi ik_1}{N_1}}$$
$$u_2^{k_2} = v_2^{k_2} = e^{\frac{2\pi i k_2}{N_2}}$$

which gives

$$\langle u^k| = \langle u_1^{k_1}\, u_2^{k_2}|,\ \langle v^k| = \langle v_1^{k_1}\, v_2^{k_2}|.$$

On identifying these vectors as the eigenvectors of sets of two commutative unitary operators, we can define $U_{1,2}$, $V_{1,2}$ by the reciprocal relations

$$\langle u_1^{k_1}\, u_2^{k_2}|\, V_1 = \langle u_1^{k_1+1}\, u_2^{k_2}|,\ \langle u_1^{k_1}\, u_2^{k_2}|\, V_2 = \langle u_1^{k_1}\, u_2^{k_2+1}|$$
$$\langle v_1^{k_1}\, v_2^{k_2}|\, U_1^{-1} = \langle v_1^{k_1+1}\, v_2^{k_2}|,\ \langle v_1^{k_1}\, v_2^{k_2}|\, U_2^{-1} = \langle v_1^{k_1}\, v_2^{k_2+1}|$$

which reproduce the properties

$$U_j^{N_j} = V_j^{N_j} = 1,\ V_j\, U_j = e^{\frac{2\pi i}{N_j}}\, U_jV_j,$$

together with the commutativity of any two operators carrying different subscripts. The orthonormality of the $N_1^2N_2^2 = N^2$ operators $X(m_1m_2n_1n_2)$ can now be directly verified. Also, by an appropriate extension of the preceeding discussion, we obtain the transformation function

$$\langle u_1^{k_1}\, u_2^{k_2}|\, v_1^{l_1}\, v_2^{l_2}\rangle = \prod_{j=1}^{2} \langle u_j^{k_j}|\, v_j^{l_j}\rangle,$$

with

$$\langle u_j^{k_j}|\, v_j^{l_j}\rangle = N_j^{-1/2}e^{\frac{2\pi i}{N_j}k_j\, l_j}.$$

The continuation of the factorization terminates in

$$N = \prod_{j=1}^{f} \nu_j,$$

where f is the total number of prime factors in N, including repetitions. We call this characteristic property of N the number of degrees of freedom for a system possessing N states. The resulting commutatively factored basis

$$X(mn) = \prod_{j=1}^{f} X(m_j n_j)$$

$$X(m_j n_j) = \nu_j^{-1/2} U_j^{m_j} V_j^{n_j}; \; m_j, n_j = 0 \,.\,.\, \nu_j - 1$$

is thus constructed from the operator bases individually associated with the f degrees of freedom, and the pair of irreducible complementary quantities of each degree of freedom is classified by the value of the prime integer $\nu = 2, 3, 5, \ldots \infty$. In particular, for $\nu = 2$ the complementary operators U and V are anticommutative and of unit square. Hence, they can be identified with σ_1 and σ_2, for example, and the operator basis is completed by the product $-iUV = \sigma_3$.

The characteristics of a degree of freedom exhibiting an infinite number of states can be investigated by making explicit the Hermitian operators upon which U and V depend,

$$U = e^{i\epsilon q}, \; V = e^{i\epsilon p}, \; \epsilon = (2\pi/\nu)^{1/2},$$

and where

$$q^k = p^k = k\epsilon, \quad k = 0, \pm 1, \pm 2, \ldots \pm {}^1\!/_2(\nu - 1).$$

We shall not carry out the necessary operations for $\nu \rightarrow \infty$, which evidently yield the well-known pair of complementary properties with continuous spectra. One remark must be made, however. In this approach one does not encounter the somewhat awkward situation in which the introduction of continuous spectra requires the construction of a new formalism, be it expressed in the language of Dirac's delta function, or of distributions. Rather, we are presented with the direct problem of finding the nature of the subspaces of physically meaningful states and operators for which the limit $\nu \rightarrow \infty$ can be performed uniformly.

* Publication assisted by the Office of Scientific Research, United States Air Force.

[1] For the notation and concepts used here see these PROCEEDINGS, **45**, 1542 (1959), and **46**, 257 (1960).

[2] The absence in the available literature of an explicit statement of this simple, general result is rather surprising. The inverse calculation giving the three-dimensional rotation matrix in terms of the elements of the unitary matrix is very well known (rotation parametrizations of Euler, Cayley-Klein), and the construction of the unitary matrix with the aid of Eulerian angles is also quite familiar.

[3] Operators having the algebraic properties of U and V have long been known from the work of Weyl, H., *Theory of Groups and Quantum Mechanics* (New York: E. P. Dutton Co., 1932), chap. 4, sect. 14, but what has been lacking is an appreciation of these operators as generators of a complete operator basis for any N, and of their optimum incompatibility, as summarized in the attribute of complementarity. Nor has it been clearly recognized that an *a priori* classification of all possible types of physical degrees of freedom emerges from these considerations.

CHAPTER THREE
THE DYNAMICAL PRINCIPLE

A measurement is a physical operation in space and in time. The properties of a system are described in relation to measurements at a given time, and no value of the time is intrinsically distinguished from another by the results of measurements on an isolated physical system.

Hence the operators symbolizing analogous properties at different times must be related by a unitary transformation. The propagation in time of the disturbance produced by a measurement implies that physical quantities referring to different times are incompatible, in general. Accordingly, complete sets of compatible properties will pertain to a common time, and the characterization of a state requires the specification of the values of these quantities (together with a spatial coordinate system) and of the time. The transformation function relating two arbitrary descriptions thus appears as $\langle a_1' t_1 | a_2'' t_2 \rangle$, special cases of which are $\langle a't | b't \rangle$, connecting two different sets of quantities at the same time, and $\langle a' t_1 | a'' t_2 \rangle$, which relates analogous properties at different times. The connection between states at two different times involves the entire dynamical history of the system in the interval. Hence the properties of specific systems must be contained completely in a dynamical principle that characterizes the general transformation function.

3.1 THE ACTION OPERATOR

Any infinitesimal alteration of the transformation function $\langle a_1' t_1 | a_2'' t_2 \rangle$ can be expressed as [Eq. (1.68)]

$$\delta \langle a_1' t_1 | a_2'' t_2 \rangle = i \langle a_1' t_1 | \delta W_{12} | a_2'' t_2 \rangle \qquad (3.1)$$

where δW_{12} is an infinitesimal Hermitian operator with the additivity property

$$\delta W_{12} + \delta W_{23} = \delta W_{13} \qquad (3.2)$$

for consecutive transformations. We now state our fundamental dynamical postulate: There exists a special class of infinitesimal alterations for which the associated operators δW_{12} are obtained by appropriate variation of a single operator, the action operator W_{12},

$$\delta W_{12} = \delta [W_{12}] \quad . \qquad (3.3)$$

It is consistent with the properties of the infinitesimal operators to assert that the action operator is Hermitian,

$$W_{12}^{\dagger} = W_{12} \quad , \qquad (3.4)$$

and has the additive combinatorial property

$$W_{12} + W_{23} = W_{13} \quad . \qquad (3.5)$$

3.2 LAGRANGIAN OPERATOR

If one views the transformation from the $a_2 t_2$ description to the $a_1 t_1$ description as occurring continuously in time through an infinite succession of infinitesimally differing descriptions, the additivity property of action operators asserts that

$$W_{12} = \sum_{t_2}^{t_1} W_{t+dt,t} \quad , \tag{3.6}$$

where

$$W_{t,t} = 0 \tag{3.7}$$

since $\langle a't|a''t\rangle$ has fixed numerical values. On writing

$$W_{t+dt,t} = dt\, L(t) \quad , \tag{3.8}$$

the action operator acquires the general form

$$W_{12} = \int_{t_2}^{t_1} dt\, L \tag{3.9}$$

in which $L(t)$, the Lagrangian operator, is a Hermitian function of some fundamental dynamical variables $x_a(t)$ in the infinitesimal neighborhood of time t. There is no loss of generality in taking the operators $x_a(t)$ to be Hermitian, and, for our present purposes, we suppose their number to be finite. The conceivable objects of variation in the action operator are the terminal times t_1 and t_2, the dynamical variables, and the structure of the Lagrangian operator.

3.3 STATIONARY ACTION PRINCIPLE

For a given dynamical system, changes in a transformation function can be produced only by explicit alteration of the states to which it refers. Such variations of states arise from changes of the physical properties or of the time involved in the definition of state, and infinitesimal eigenvector transformations are generated by Hermitian operators that depend only upon dynamical variables at the stated time,

$$\begin{aligned} \delta\langle a_1' t_1| &= i\langle a_1' t_1|G_1 \\ \delta|a_2'' t_2\rangle &= -\, iG_2|a_2'' t_2\rangle\ . \end{aligned} \tag{3.10}$$

Hence,

$$\delta\langle a_1' t_1 | a_2'' t_2\rangle = i\langle a_1' t_1 | (G_1 - G_2) | a_2'' t_2\rangle \quad , \qquad (3.11)$$

and, with a given dynamical system,

$$\delta W_{12} = \delta\left[\int_{t_2}^{t_1} dt\, L(t) \right] = G_1 - G_2 \quad , \qquad (3.12)$$

which is the operator principle of stationary action for it asserts that the variation of the action integral involves only dynamical variables at the terminal times. This principle implies equations of motion for the dynamical variables and yields specific forms for generators of infinitesimal transformations. Note that the Lagrangian operator cannot be determined completely by the dynamical nature of the system, since we must be able to produce a variety of infinitesimal transformations for a given system. Indeed, if two Lagrangians differ by a time derivative,

$$\bar{L} = L - \frac{d}{dt} W \quad , \qquad (3.13)$$

the action operators are related by

$$\bar{W}_{12} = W_{12} - (W_1 - W_2) \tag{3.14}$$

and

$$\begin{aligned}\delta\bar{W}_{12} &= \delta W_{12} - (\delta W_1 - \delta W_2) \\ &= \bar{G}_1 - \bar{G}_2 \quad ,\end{aligned} \tag{3.15}$$

which also satisfies the stationary action requirement but implies new generators of infinitesimal transformations at times t_1 and t_2, as given by

$$\delta W_1 = G_1 - \bar{G}_1 \quad , \quad \delta W_2 = G_2 - \bar{G}_2 \tag{3.16}$$

3.4 THE HAMILTONIAN OPERATOR

The development of the fundamental dynamical variables through an infinitesimal time interval is described by an infinitesimal unitary transformation, which implies first order differential equations of motion for these variables. A general form of Lagrangian operator that yields first order equations of motion is

$$L = \tfrac{1}{4} \sum_{ab} \left(x_a A_{ab} \frac{dx_b}{dt} - \frac{dx_a}{dt} A_{ab} x_b\right) - H(x_a, t)$$

$$= \tfrac{1}{4} \left(xA \frac{dx}{dt} - \frac{dx}{dt} Ax\right) - H(x, t), \qquad (3.17)$$

in which A_{ab} is a numerical matrix. We shall speak of the two parts of the Lagrangian operator as the kinematical and the dynamical parts, respectively. Note that the kinematical part has been symmetrized with respect to transference of the time derivative. The Hermitian requirement on L applies to the two parts separately. Hence H, the Hamiltonian operator, must be Hermitian, and the finite matrix A must be skew - Hermitian.

$$A^{T*} = -A . \qquad (3.18)$$

For an isolated dynamical system, there is no explicit reference to time in H.

3.5 EQUATIONS OF MOTION. GENERATORS

The action operator implied by the particular structure of L in (3.17) is

$$W_{12} = \int_{t_2}^{t_1} [\tfrac{1}{4}(xA\, dx - dx\, Ax) - H\, dt] . \qquad (3.19)$$

In this form the limits of integration are objects of variation. However, one can think of introducing an auxiliary variable τ, and producing the variations δt_1, δt_2 by altering the functional dependence of $t = t(\tau)$ upon τ, with fixed limits τ_1, τ_2. This procedure places the time variable on somewhat the same footing as the dynamical variables. Now

$$\delta[\tfrac{1}{4}(xA\,dx - dx\,Ax) - H\,dt] - d[\tfrac{1}{4}(xA\,\delta x - \delta x\,Ax) - H\,\delta t] = \tfrac{1}{2}(\delta x\,A\,dx - dx\,A\,\delta x) - \delta H\,dt + dH\,\delta t \qquad (3.20)$$

since

$$\begin{aligned} \delta dx &= \delta[x(\tau+d\tau) - x(\tau)] = d\delta x\,, \\ \delta dt &= \delta[t(\tau+d\tau) - t(\tau)] = d\delta\tau\,, \end{aligned} \qquad (3.21)$$

and thus the stationary action principle asserts that

$$\tfrac{1}{2}(\delta x\,A\,dx - dx\,A\,\delta x) = \delta H\,dt - dH\,\delta t\,, \qquad (3.22)$$

or

$$\delta H = \frac{dH}{dt}\,\delta t + \tfrac{1}{2}\left(\delta x\,A\,\frac{dx}{dt} - \frac{dx}{dt}\,A\,\delta x\right), \qquad (3.23)$$

and yields the infinitesimal generators

$$G = \tfrac{1}{4}(xA\,\delta x - \delta x\,Ax) - H\,\delta t$$
$$= G_x + G_t \tag{3.24}$$

at the terminal times.

The structure of the Hamiltonian operator is as yet unspecified. If its variation is to possess the form (3.23), with the δx appearing only on the left and on the right, these variations must possess elementary operator properties characterizing the special class of operator variations to which the dynamical principle refers. Thus, we should be able to displace each δx_a entirely to the left or to the right, in the structure of δH, which defines the left and the right derivatives of H,

$$\delta H - \frac{\partial H}{\partial t}\,\delta t = \delta x\,\frac{\partial_\ell H}{\partial x} = \frac{\partial_r H}{\partial x}\,\delta x \quad . \tag{3.25}$$

In view of the complete symmetry between left and right, we infer that the terms in (3.23) with δx on the left, and on the right, are identical. We thus obtain

$$\frac{dH}{dt} = \frac{\partial H}{\partial t} \tag{3.26}$$

and

$$\begin{aligned} A \frac{dx}{dt} &= \frac{\partial_\ell H}{\partial x} \quad , \\ - \frac{dx}{dt} A = - A^T \frac{dx}{dt} &= \frac{\partial_r H}{\partial x} \quad , \end{aligned} \tag{3.27}$$

where the latter must be equivalent forms of the equations of motion. Similarly, the infinitesimal generator G_x possesses the equivalent forms

$$\begin{aligned} G_x &= - \tfrac{1}{2} \, \delta x \, A \, x \\ &= \tfrac{1}{2} \, x \, A \, \delta x = \tfrac{1}{2} \, (A^T x) \, \delta x \quad , \end{aligned} \tag{3.28}$$

and, generally, left and right hand forms are connected by the adjoint operation. We shall assume in our discussion that the matrix A is nonsingular. This implies that every variation δx_a appears independently in G_x and that each variable x_a obeys an explicit equation of motion.

3.6 COMMUTATION RELATIONS

The infinitesimal generator $G_t = - H \, \delta t$

evidently describes the unitary transformation from a description at time t to the analogous one at time $t + \delta t$, which identifies H as the energy operator of the system. If F is any function of the dynamical variables $x(t)$ and of the time, the operator $\bar{F}$ that plays the role of F in the description referring to time $t + \delta t$ is

$$\bar{F} = F - \delta F = F\big(x(t+\delta t)\ ,\ t\big) \tag{3.29}$$

since the numerical parameter t is not affected by operator transformations. Hence

$$\frac{1}{i}\,[F\ ,\ G_t] = \delta F = -\left\{\frac{dF}{dt} - \frac{\partial F}{\partial t}\right\}\delta t \tag{3.30}$$

or

$$\frac{dF}{dt} = \frac{\partial F}{\partial t} + \frac{1}{i}\,[F\ ,\ H]\ , \tag{3.31}$$

which is the general equation of motion. On placing $F = H$, we obtain

$$\frac{dH}{dt} = \frac{\partial H}{\partial t}\ , \tag{3.32}$$

in agreement with (3.26), derived from the stationary action principle. Such consistency must also appear in the equations of motion for the dynamical variables, which requires that

$$\frac{1}{i}\,[Ax\ ,\ H] = \frac{\partial_\ell H}{\partial x} \tag{3.33}$$

and

$$\frac{1}{i}\,[-xA\ ,\ H] = \frac{\partial_r H}{\partial x} \quad . \tag{3.34}$$

On writing these relations in the form

$$\begin{aligned} \frac{1}{i}\,\delta x\ [Ax\ ,\ H] &= \delta x\ \frac{\partial_\ell H}{\partial x} \\ \frac{1}{i}\,[-xA\ ,\ H]\ \delta x &= \frac{\partial_r H}{\partial x}\,\delta x \quad , \end{aligned} \tag{3.35}$$

we can conclude from the equality of the right sides, and from the two equivalent expressions for G_x , that

$$[\delta x\ ,\ H]\ Ax = -\ xA[\delta x\ ,\ H] \quad . \tag{3.36}$$

We shall satisfy this consistency requirement by demanding that each variation δx_a commute with the Hamiltonian operator,

$$[\delta x_a\ ,\ H] = 0\ , \tag{3.37}$$

for this enables the commutation relations (3.35) to be stated as

$$\frac{1}{i} [H , G_x] = \tfrac{1}{2}\, \delta x \frac{\partial_\ell H}{\partial x} = \frac{\partial_r H}{\partial x} \tfrac{1}{2}\, \delta x \qquad (3.38)$$

and identifies the transformation generated by G_x as the change of each dynamical variable x_a by $\frac{1}{2}\delta x_a$. Accordingly, for an arbitrary function of the dynamical variables at the stated time, we have

$$\frac{1}{i} [F , G_x] = \tfrac{1}{2}\, \delta x \frac{\partial_\ell H}{\partial x} = \frac{\partial_r H}{\partial x} \tfrac{1}{2}\, \delta x \quad . \qquad (3.39)$$

3.7 THE TWO CLASSES OF DYNAMIC VARIABLES

The two equivalent versions of G_x , bilinear in x and δx , indicate that displacing a variation δx_a across any of the dynamical variables induces a linear transformation on these variables,

$$\delta x_a \, x_b = (k_a x)_b \, \delta x_a \quad , \qquad (3.40)$$

where k_a designates a matrix. The adjoint statement is

$$x_b \, \delta x_a = \delta x_a \left(k_a^* x\right)_b \quad , \tag{3.41}$$

from which we conclude that

$$k_a^* k_a = 1 \ . \tag{3.42}$$

The commutativity of each δx_a with the Hamiltonian operator $H(x)$ now appears as the set of invariance properties

$$\begin{aligned} H(k_a x) &= H(x) \\ &= H\left(k_a^{-1} x\right) \quad , \end{aligned} \tag{3.43}$$

where the latter statement expresses the Hermitian nature of H . On forming the commutator with a second variation δx_b we learn that

$$H(k_a k_b x) = H(x) \ , \tag{3.44}$$

which shows that the set of linear transformations k_a form a group of invariance transformations for H . From the fundamental significance of this group we conclude that it must apply to the complete structure of the Lagrangian operator. The kinematical term is invariant under the linear transformation k_a if

$$k_a^T A k_a = A \, . \tag{3.45}$$

Also, the two equivalent expressions for G_x imply that

$$\begin{aligned}(Ak_a)_{ab} &= - A_{ba} \\ &= (k_a^{T^{-1}} A)_{ab} \quad ,\end{aligned} \tag{3.46}$$

which is consistent with (3.45). For this intrinsic equivalence of the variables x and $k_a x$ to be complete, the latter must be Hermitian operators. Hence the matrices k_a are real, and obey

$$k_a{}^2 = 1 \, . \tag{3.47}$$

The construction of the invariance group described by the matrices k_a has been based upon the particular operator variation δx_a. But there must exist some freedom of linear transformation whereby new Hermitian variables are introduced,

$$\bar{x} = \ell x \ , \tag{3.48}$$

with accompanying redefinitions of the matrix A and of the Hamiltonian operator.

$$\bar{A} = \ell^{T^{-1}} A \ell^{-1} \ , \qquad \bar{H}(\bar{x}) = H(x) = H\left(\ell^{-1} \bar{x}\right) \ . \tag{3.49}$$

For the new choice of variables, the invariance properties of the Hamiltonian read

$$\bar{H}(\bar{x}) = H(k_a x) = \bar{H}\left(\ell \ k_a \ \ell^{-1} \ \bar{x}\right) \ , \tag{3.50}$$

so that

$$\bar{k}_a = \ell \ k_a \ \ell^{-1} \ . \tag{3.51}$$

On the other hand, these matrices should appear directly from the commutation properties of the variations $\delta\bar{x}_a$,

$$\delta\bar{x}_a \ \bar{x}_b = \left(\bar{k}_a \ \bar{x}\right)_b \ \delta\bar{x}_a \ . \tag{3.52}$$

By referring this statement back to the characteristics of the δx , we obtain

$$\sum_b \ell_{ab} k_b (\ell^{-1})_{bc} = \delta_{ac} \ell^{-1} \bar{k}_a \ell$$
$$= \delta_{ac} k_a \quad . \tag{3.53}$$

Now this result cannot be valid for arbitrary ℓ, within a certain group of transformations, unless the matrices k_b are identical for all values of b that can be connected by the linear transformation. Accordingly, the dynamical variables x must decompose into classes such that linear transformations within each class only are permissible, with the different classes distinguished by specific matrices k . In view of the freedom of independent linear transformations within each class, the matrices k_a must maintain the decomposition into classes and thus contain only submatrices characteristic of each class of variable. The partitioning of the dynamical variables produced by any of the k_a matrices,

$$x = \tfrac{1}{2}(1 + k_a)x + \tfrac{1}{2}(1 - k_a)x \quad , \tag{3.54}$$

combined with the variation properties

$$\delta x_a \tfrac{1}{2}(1 \pm k_a)x = \pm \tfrac{1}{2}(1 \pm k_a)x \, \delta x_a \quad , \tag{3.55}$$

indicate further that the submatrices of k_a appear merely as numbers k_{ab} labelling the various classes. According to (3.47), these numbers obey

$$(k_{ab})^2 = 1 . \tag{3.56}$$

Hence the two possibilities of $k_{aa} = \pm 1$ define two distinct classes of dynamical variables,

$$k_{11} = +1 , \quad k_{22} = -1 . \tag{3.57}$$

We shall see shortly that $k_{12} = k_{21}$, and, since the identity transformation must appear in the group of k-transformations, we have

$$k_{12} = +1 , \quad k_{21} = +1 . \tag{3.58}$$

If we distinguish the two classes of dynamical variables as variables of the first kind, z_k, and variables of the second kind, ζ_κ, the operator properties of the variations δz_k, $\delta\zeta_\kappa$, summarized in

$$\delta x_a \, x_b = k_{ab} x_b \, \delta x_a , \tag{3.59}$$

are given explicitly by

$$[\delta z_k , z_\ell] = 0 , \quad [\delta z_k , \zeta_\kappa] = 0$$
$$[\delta\zeta_\kappa , z_k] = 0 , \quad \{\delta\zeta_\kappa , \zeta_\lambda\} = 0 , \tag{3.60}$$

where the curly bracket signifies the anticommutator

$$\{A , B\} = AB + BA . \tag{3.61}$$

The Hamiltonian invariance transformation implied by commutativity with the $\delta\zeta$,

$$H(z , -\zeta) = H(z , \zeta) , \tag{3.62}$$

requires that H be an even function of the variables of the second kind, but no restriction appears for the dependence upon the variables of the first kind.

The property of the matrix A contained in (3.45), together with the opposite signs of k_{22} and k_{21} , shows that all elements of A connecting the two classes of variables must be zero. Hence A reduces completely into two submatrices associated with the two kinds of variable, which we shall designate as a and $i\alpha$, respectively.

It follows from (3.46) that a is antisymmetrical, and hence real,

$$a = -a^T = a^* , \tag{3.63}$$

while α is symmetrical and real,

$$\alpha = \alpha^T = \alpha^* . \tag{3.64}$$

Two complete reduction of A into a and $i\alpha$ implies a corresponding additive decomposition of the generator G_x ,

$$G_x = G_z + G_\zeta , \tag{3.65}$$

with

$$G_z = \tfrac{1}{2} za\, \delta z = \tfrac{1}{2}(a\, \delta z)z \tag{3.66}$$

and

$$G_\zeta = \tfrac{1}{2}\zeta i\alpha\, \delta\zeta = -\tfrac{1}{2}(i\alpha\, \delta\zeta)\zeta . \tag{3.67}$$

This structure of G_x is an aspect of the additive form assumed by the kinematical term in the Lagrangian,

$$L = \tfrac{1}{4}\left(xA\,\frac{dx}{dt} - \left(A^T\,\frac{dx}{dt}\right)x\right) - H$$
$$= \tfrac{1}{4}\{z\ ,\ a\,\frac{dz}{dt}\} + \tfrac{1}{4}[\zeta\ ,\ i\alpha\,\frac{d\zeta}{dt}] - H\ , \qquad (3.68)$$

which we express by calling the two sets of variables kinematically independent. The equations of motion for these kinematically independent sets of variables are

$$a\,\frac{dz}{dt} = \frac{\partial_\ell H}{\partial z} = \frac{\partial_r H}{\partial z} \qquad (3.69)$$

and

$$i\alpha\,\frac{d\zeta}{dt} = \frac{\partial_\ell H}{\partial \zeta} = -\,\frac{\partial_r H}{\partial \zeta}\ . \qquad (3.70)$$

We shall adopt a uniform notation to indicate the characteristic symmetrization in the Lagrangian of bilinear structures referring to variables of the first kind, and the antisymmetrization for bilinear functions of variables of the second kind, namely

$$L = \tfrac{1}{2}z\ .\ a\,\frac{dz}{dt} + \tfrac{1}{2}\zeta\ .\ i\alpha\,\frac{d\zeta}{dt} - H\ . \qquad (3.71)$$

On displacing δx to the left or to the right in the general commutation relation (3.39), we obtain

$$(Ax)_a \; F(x) \; - \; F(k_a x) \; (Ax)_a = i \frac{\partial_\ell F}{\partial x_a}$$
$$F(x) \; (xA)_a \; - \; (xA)_a \; F(k_a x) = i \frac{\partial_r F}{\partial x_a} \quad . \tag{3.72}$$

The equations resulting from the special choice $F(x) = x_b$ can be presented as

$$x_a x_b - k_{ab} x_b x_a = i(A^{-1})_{ab}$$
$$x_b x_a - k_{ab} x_a x_b = i(A^{-1})_{ba} \quad , \tag{3.73}$$

and, on interchanging a and b in one version, we conclude from the other that

$$k_{ab} = k_{ba} \quad . \tag{3.74}$$

We have already made use of the only significant statement contained here, $k_{12} = k_{21}$. The commutation properties of the two classes of dynamical variables now appear explicitly as

$$[z_k , z_\ell] = i(a^{-1})_{k\ell}$$

$$\{\zeta_\kappa , \zeta_\lambda\} = (\alpha^{-1})_{\kappa\lambda} \qquad (3.75)$$

$$[z_k , \zeta_\kappa] = 0 \quad .$$

It will be seen that the structure of the operators reproduces that of the matrices, with the antisymmetrical, skew-Hermitian commutator appearing with the antisymmetrical, imaginary matrix $(1/i)a$, while the symmetrical, Hermitian anticommutator is related to the syymetrical, real matrix α . In addition, the matrix α must be positive-definite if the variables ζ_κ are to be linearly independent. The explicit forms of the general commutation relation are

$$\frac{1}{i} [az , F] = \frac{\partial F}{\partial z} \quad , \qquad (3.76)$$

without distinction between left and right derivatives, while, by distinguishing operators that are even and odd functions of the variables of the second kind, we find that

$$[\alpha\zeta , F_e] = \frac{\partial_\ell F_e}{\partial \zeta} = - \frac{\partial_r F_e}{\partial \zeta} \qquad (3.77)$$

and

$$\{\alpha \zeta \;,\; F_o\} = \frac{\partial_\ell F_o}{\partial \zeta} = \frac{\partial_r F_o}{\partial \zeta} \quad . \tag{3.78}$$

3.8 COMPLEMENTARY VARIABLES OF THE FIRST KIND

The matrices a and α are nonsingular, according to the assumption made about A . For the antisymmetrical matrix a , associated with variables of the first kind, the remark that

$$\det a = \det a^T = \det (-a) \tag{3.79}$$

shows that the number of variables of the first kind cannot be odd. We shall designate this number as $2n_1$. Now the matrix defined by

$$\mathit{a} = (a^T a)^{\frac{1}{2}} \tag{3.80}$$

is a real, symmetrical, positive-definite function of a , and on writing

$$a = \mathit{a}\lambda = \lambda \mathit{a} \;, \tag{3.81}$$

we observe that λ is a real, antisymmetrical matrix that obeys

$$\lambda^2 = -1 \quad . \tag{3.82}$$

Furthermore, there exists a real, symmetrical matrix ρ such that

$$\rho a \rho^{-1} = -a , \quad \rho^2 = 1 . \tag{3.83}$$

The matrix ρ commutes with α and anticommutes with λ. By an appropriate choice of Hermitian variables z, all these $2n_1$-dimensional matrices can be displayed in a partitioned form, with n_1-dimensional submatrices, as

$$\rho = \begin{pmatrix} 1 & 0 \\ 0 & -1 \end{pmatrix} , \quad \lambda = \begin{pmatrix} 0 & 1 \\ -1 & 0 \end{pmatrix} , \quad \alpha = \begin{pmatrix} \alpha & 0 \\ 0 & \alpha \end{pmatrix} , \tag{3.84}$$

which establishes the possibility of attaining the matrix forms

$$a = \begin{pmatrix} 0 & \alpha \\ -\alpha & 0 \end{pmatrix} , \quad a^{-1} = \begin{pmatrix} 0 & -\alpha^{-1} \\ \alpha^{-1} & 0 \end{pmatrix} , \tag{3.85}$$

with α an n_1-dimensional real, symmetrical, positive-definite matrix. We shall call the correspondingly partitioned sets of n_1 variables $z_k^{(1)}$ and $z_k^{(2)}$, $k = 1,\ldots,n_1$. According to (3.75), these variables obey the commutation relations

$$[z_k^{(1)}, z_\ell^{(1)}] = [z_k^{(2)}, z_\ell^{(2)}] = 0$$
$$[z_k^{(1)}, z_\ell^{(2)}] = -i(a^{-1})_{k\ell} \quad . \qquad (3.86)$$

With the partitioned form of a, the kinematical term and the infinitesimal generator, referring to the variables of the first kind, acquire the forms

$$\tfrac{1}{2}\left(z^{(1)} \cdot a \frac{dz^{(2)}}{dt} - \frac{dz^{(1)}}{dt} \cdot a\, z^{(2)}\right) \qquad (3.87)$$

and

$$G_z = \tfrac{1}{2}\left(z^{(1)} a\, \delta z^{(2)} - \delta z^{(1)} a\, z^{(2)}\right) , \qquad (3.88)$$

where the latter generates changes in $z^{(1)}$ and $z^{(2)}$ of $\tfrac{1}{2}\delta z^{(1)}$ and $\tfrac{1}{2}\delta z^{(2)}$, respectively. Now let us exploit the freedom to add a time derivative term to the Lagrangian with a corresponding alteration of the infinitesimal generator, in the manner of (3.13) and (3.16). With the choice

$$W = -\tfrac{1}{2}\, z^{(1)} \cdot a\, z^{(2)} \qquad (3.89)$$

we obtain the new kinematical term

$$z^{(1)} \cdot a \frac{dz^{(2)}}{dt} \tag{3.90}$$

and the new generator

$$G_{z^{(2)}} = z^{(1)} a \delta z^{(2)} , \tag{3.91}$$

while the opposite sign for W gives the kinematical term

$$- \frac{dz^{(1)}}{dt} \cdot a z^{(2)} = - z^{(2)} \cdot a \frac{dz^{(1)}}{dt} \tag{3.92}$$

and the generator

$$G_{z^{(1)}} = - \delta z^{(1)} a z^{(2)} = - z^{(2)} a \delta z^{(1)} . \tag{3.93}$$

By comparison with (3.88) we recognize that $G_{z^{(2)}}$ generates the transformation in which the operators $z^{(1)}$ are unaltered and the $z^{(2)}$ are changed by $\delta z^{(2)}$. The converse interpretation applies to $G_{z^{(1)}}$. Thus we have divided the variables of the first kind into two sets that are complementary, each set comprising the generators of infinitesimal variations of the other set. The interpretation of the generators $G_{z^{(1)}}$ and $G_{z^{(2)}}$ is expressed by

$$\frac{1}{i}\,[az^{(2)}\,,\,F] = \frac{\partial F}{\partial z^{(1)}}$$

$$\frac{1}{i}\,[F\,,\,az^{(1)}] = \frac{\partial F}{\partial z^{(2)}}\,, \tag{3.94}$$

from which we regain the commutation relations (3.86), and derive the equations of motion for the complementary variables,

$$a\,\frac{dz^{(2)}}{dt} = \frac{\partial H}{\partial z^{(1)}}\,, \qquad -\,a\,\frac{dz^{(1)}}{dt} = \frac{\partial H}{\partial z^{(2)}}\,, \tag{3.95}$$

in agreement with the implications of the action principle.

A real, symmetrical positive-definite matrix can always be reduced to the unit matrix by a real transformation, which, applied to a, places the description by complementary variables in a canonical form. The required transformation is produced by introducing the canonical variables

$$p = a^{\frac{1}{2}}\,z^{(1)}\,, \qquad q = a^{\frac{1}{2}}\,z^{(2)} \tag{3.96}$$

for this converts the kinematical term (3.90) into

$$p \cdot \frac{dq}{dt} = \sum_{k=1}^{n_1} p_k \cdot \frac{dq_k}{dt} \quad . \tag{3.97}$$

Thus the canonical variables constitute n_1 complementary pairs of kinematically independent dynamical variables of the first kind. Various aspects of the canonical variables are: the equations of motion,

$$\frac{dq_k}{dt} = \frac{\partial H}{\partial p_k} \quad , \quad -\frac{dp_k}{dt} = \frac{\partial H}{\partial q_k} \quad k = 1,\ldots,n_1 \ , \tag{3.98}$$

the infinitesimal generators,

$$G_q = \sum_k p_k \, \delta q_k \quad , \quad G_p = -\sum_k \delta p_k \, q_k \quad , \tag{3.99}$$

the general commutation relations,

$$\frac{1}{i} [F \, , \, p_k] = \frac{\partial F}{\partial q_k} \quad , \quad \frac{1}{i} [q_k \, , \, F] = \frac{\partial F}{\partial p_k} \quad , \tag{3.100}$$

and the commutation properties of the canonical variables,

$$\begin{aligned} [q_k \, , \, q_\ell] &= [p_k \, , \, p_\ell] = 0 \\ [q_k \, , \, p_\ell] &= i \, \delta_{k\ell} \quad . \end{aligned} \tag{3.101}$$

3.9 NON-HERMITIAN VARIABLES OF THE FIRST KIND

It is important to recognize that the dynamical theory, which has been developed in terms of Hermitian dynamical variables, permits the introduction of non-Hermitian complementary variables. Let us define

$$y = 2^{-\frac{1}{2}}\left(z^{(2)}+iz^{(1)}\right) , \quad iy^{\dagger} = 2^{-\frac{1}{2}}\left(z^{(1)}+iz^{(2)}\right) \tag{3.102}$$

and observe that the kinematical term (3.87) can also be written as

$$\tfrac{1}{2}\left(y^{\dagger} \cdot ia \frac{dy}{dt} - \frac{dy^{\dagger}}{dt} \cdot ia\, y\right) \tag{3.103}$$

which still has the same structure, with $iy^{\dagger}$ and y replacing $z^{(1)}$ and $z^{(2)}$, respectively. The latter form also persists under arbitrary complex linear transformations of the non-Hermitian variables y , with appropriate redefinitions of a as a positive-definite Hermitian matrix. The formal application of the previous considerations to the non-Hermitian variables now leads, for example, to the commutation relations

$$[\alpha y \, , \, F] = \frac{\partial F}{\partial y^{\dagger}}$$

$$[F \, , \, \alpha y^{\dagger}] = \frac{\partial F}{\partial y} \; , \qquad (3.104)$$

written for real α , which are precisely the results that would be obtained by combining the commutation relations for the Hermitian variables, in accordance with the definitions of the non-Hermitian variables, together with

$$\frac{\partial F}{\partial y} = 2^{-\frac{1}{2}}\left(\frac{\partial F}{\partial z^{(2)}} - i \, \frac{\partial F}{\partial z^{(1)}} \right)$$

$$-i \, \frac{\partial F}{\partial y^{\dagger}} = 2^{-\frac{1}{2}}\left(\frac{\partial F}{\partial z^{(1)}} - i \, \frac{\partial F}{\partial z^{(2)}} \right) \; . \qquad (3.105)$$

Regarded as the definition of differentiation with respect to the non-Hermitian variables, these equations imply that

$$\frac{\partial y_k{}^{\dagger}}{\partial y_\ell} = \frac{\partial y_k}{\partial y_\ell{}^{\dagger}} = 0 \; ,$$

$$\frac{\partial y_k}{\partial y_\ell} = \frac{\partial y_k{}^{\dagger}}{\partial y_\ell{}^{\dagger}} = \delta_{k\ell} \; , \qquad (3.106)$$

which justifies the formal treatment in which y

and $y^\dagger$ are subjected to independent variations. Thus the formal theory employing non-Hermitian variables produces correct equations of motion and commutation relations. In particular, the canonical version is applicable, although the canonical variables q_k and p_k are not self-adjoint but rather

$$p_k = iq_k^\dagger \quad . \tag{3.107}$$

Accordingly, the canonical equations of motion and commutation relations can be written

$$i\,\frac{dq_k}{dt} = \frac{\partial H}{\partial q_k{}^\dagger} \quad , \qquad -i\,\frac{dq_k{}^\dagger}{dt} = \frac{\partial H}{\partial q_k} \quad , \tag{3.108}$$

and

$$\begin{aligned} [q_k \,,\, q_\ell] &= [q_k{}^\dagger \,,\, q_\ell{}^\dagger] = 0 \\ [q_k \,,\, q_\ell{}^\dagger] &= \delta_{k\ell} \quad , \end{aligned} \tag{3.109}$$

in which the pairs of equations stand in adjoint relation.

3.10 COMPLEMENTARY VARIABLES OF THE SECOND KIND

The necessity of an even number of variables and the possibility of dividing them into two complementary sets also appears for variables of the second kind. Let us observe that the real, symmetrical, positive-definite matrix α can be reduced to the unit matrix by an appropriate real transformation of the variables, which is effectively produced on introducing

$$\xi = \alpha^{\frac{1}{2}}\zeta \ . \tag{3.110}$$

The new canonical Hermitian variables obey

$$\{\xi_\kappa \ , \ \xi_\lambda\} = \delta_{\kappa\lambda} \qquad \kappa,\lambda = 1,\ldots,\nu \quad , \tag{3.111}$$

which means that two different ξ -operators anticommute, and that the square of each ξ is a common numerical multiple of the unit operator. The accompanying canonical form of the equations of motion is

$$i\,\frac{d\xi}{dt} = \frac{\partial_\ell H}{\partial\xi} = -\,\frac{\partial_r H}{\partial\xi} \quad , \tag{3.112}$$

while the general commutation relations become

$$[\xi\ ,\ F_e] = \frac{\partial_\ell F_e}{\partial \xi} = -\ \frac{\partial_r F_e}{\partial \xi}\ ,$$
$$\{\xi\ ,\ F_o\} = \frac{\partial_\ell F_o}{\partial \xi} = \frac{\partial_r F_o}{\partial \xi}\ . \qquad (3.113)$$

We now want to emphasize, for a system described by variables of the second kind, the requirement that the complete measurement algebra be derived from the fundamental dynamical variables, which is the assumption that accompanies the introduction of such variables. The linearly independent operators that can be constructed from the ξ-variables are enumerated as follows: the unit operator; the ν operators ξ_κ ; the $\frac{1}{2}\nu(\nu-1)$ operators $\xi_\kappa\xi_\lambda$, $\kappa<\lambda$; the $\frac{1}{6}\nu(\nu-1)(\nu-2)$ operators $\xi_\kappa\xi_\lambda\xi_\mu$, $\kappa<\lambda<\mu$; and so forth, culminating in the single operator $\xi_1\xi_2 ,\dots, \xi_\nu$. The total number of independent operators so obtained, the dimensionality of the ξ-algebra, is

$$\sum_{\alpha=0}^{\nu} \frac{\nu!}{\alpha!(\nu-\alpha)!} = 2^\nu \ . \qquad (3.114)$$

But this number must also be the dimensionality of the measurement algebra, which equals the square of the integer N representing the total number of states. The desired equivalence is possible only if ν is an even integer,

$$\nu = 2n_2 \quad , \quad N = 2^{n_2} \quad . \tag{3.115}$$

On dividing the Hermitian canonical variables into two sets of equal number, $\xi^{(1)}$ and $\xi^{(2)}$, the kinematical term for the variables of the second kind becomes

$$\frac{i}{2}\left(\xi^{(1)} \cdot \frac{d\xi^{(1)}}{dt} + \xi^{(2)} \cdot \frac{d\xi^{(2)}}{dt}\right) \quad . \tag{3.116}$$

A description employing complementary variables appears with the introduction of non-Hermitian canonical variables of the second kind,

$$\begin{aligned} q &= 2^{-\frac{1}{2}}\left(\xi^{(2)} + i\xi^{(1)}\right) \\ p &= iq^{\dagger} = 2^{-\frac{1}{2}}\left(\xi^{(1)} + i\xi^{(2)}\right) \quad , \end{aligned} \tag{3.117}$$

which converts the kinematical term into

$$\tfrac{1}{2}\left(p \cdot \frac{dq}{dt} - \frac{dp}{dt} \cdot q\right) \quad . \tag{3.118}$$

This structure is applicable to either of the two kinds of complementary dynamical variable, as are the forms

$$p \cdot \frac{dq}{dt} \quad , \qquad - \frac{dp}{dt} \cdot q \tag{3.119}$$

obtained by adding suitable time derivatives. Accordingly, the equation of motion

$$\frac{dq}{dt} = \frac{\partial_\ell H}{\partial p} \quad , \quad - \frac{dp}{dt} = \frac{\partial_r H}{\partial q} \tag{3.120}$$

and the generators of infinitesimal changes in q or p ,

$$G_q = p\,\delta q \quad , \quad G_p = -\,\delta p\, q \quad , \tag{3.121}$$

can refer to either of the two types of variable. The distinction between the two classes is implicit in the relation of left and right derivatives, and, generally, in the operator properties of the variations δq and δp . Thus, for the variables of the second kind,

$$\begin{aligned} G_q &= \sum_{\kappa=1}^{n_2} p_\kappa\, \delta q_\kappa = - \sum_\kappa \delta q_\kappa\, p_\kappa \\ G_p &= - \sum_\kappa \delta p_\kappa\, q_\kappa = \sum_\kappa q_\kappa\, \delta p_\kappa \quad . \end{aligned} \tag{3.122}$$

The general commutation relations that express the significance of these generators are

$$\frac{1}{i}[q_\kappa , F_e] = \frac{\partial_\ell F_e}{\partial p_\kappa} = -\frac{\partial_r F_e}{\partial p_\kappa}$$

$$\frac{1}{i}[p_\kappa , F_e] = \frac{\partial_\ell F_e}{\partial q_\kappa} = -\frac{\partial_r F_e}{\partial q_\kappa} \quad , \tag{3.123}$$

together with

$$\frac{1}{i}\{q_\kappa , F_o\} = \frac{\partial_\ell F_o}{\partial p_\kappa} = \frac{\partial_r F_o}{\partial p_\kappa}$$

$$\frac{1}{i}\{p_\kappa , F_o\} = \frac{\partial_\ell F_o}{\partial q_\kappa} = \frac{\partial_r F_o}{\partial q_\kappa} \quad , \tag{3.124}$$

and these statements are identical with the results obtained directly from (3.113), the commutation properties of the Hermitian dynamical variables.

As applications of the general commutation relations, we regain the canonical equations of motion, and derive the commutation properties of the canonical non-Hermitian variables of the second kind,

$$\{q_\kappa , q_\lambda\} = \{p_\kappa , p_\lambda\} = 0$$

$$\{q_\kappa , p_\lambda\} = i\,\delta_{\kappa\lambda} \quad . \tag{3.125}$$

In virtue of the adjoint connection between the canonical variables,

$$p_\kappa = i\, q_\kappa^\dagger \quad , \tag{3.126}$$

the equations of motion and commutation relations can also be presented as

$$i\,\frac{dq_\kappa}{dt} = \frac{\partial_\ell H}{\partial q_\kappa^\dagger} \qquad -i\,\frac{dq_\kappa^\dagger}{dt} = \frac{\partial_r H}{\partial q_\kappa} \quad , \tag{3.127}$$

and

$$\begin{aligned} \{q_\kappa\,,\, q_\lambda\} &= \{q_\kappa^\dagger\,,\, q_\lambda^\dagger\} = 0 \\ \{q_\kappa\,,\, q_\lambda^\dagger\} &= \delta_{\kappa\lambda} \quad . \end{aligned} \tag{3.128}$$

CHAPTER FOUR

THE SPECIAL CANONICAL GROUP

The commutation properties of the infinitesimal operator variations that are employed in the fundamental dynamical principle are such as to maintain the commutation relations obeyed by the dynamical variables. Accordingly, these special variations possess a transformation group aspect, which we now proceed to examine.

I. VARIABLES OF THE FIRST KIND

For variables of the first kind, the variations δz commute with all operators and thus appear as arbitrary infinitesimal real numbers. If one considers the generators of two independent infinitesimal variations

$$G_z^{(1)} = -\tfrac{1}{2}\,\delta^{(1)}z\;az \quad , \quad G_z^{(2)} = \tfrac{1}{2}za\;\delta^{(2)}z \quad , \tag{4.1}$$

their commutator can be evaluated by enploying the generator significance of either operator,

$$\frac{1}{i}\,[G_z^{(1)}\,,\,G_z^{(2)}] = \frac{\partial G_z^{(1)}}{\partial z}\,\tfrac{1}{2}\,\delta^{(2)}z = -\tfrac{1}{2}\,\delta^{(1)}z\,\frac{\partial G_z^{(2)}}{\partial z}$$
$$= -\tfrac{1}{4}\,\delta^{(1)}z\;a\;\delta^{(2)}z \quad . \tag{4.2}$$

The equivalent canonical forms are

$$[G_q^{(1)}, G_q^{(2)}] = [G_p^{(1)}, G_p^{(2)}] = 0$$
$$[G_p^{(1)}, G_q^{(2)}] = -i\,\delta^{(1)}p\,\delta^{(2)}q \quad . \tag{4.3}$$

We thus recognize that the totality of infinitesimal generators G_z, or G_q, G_p, together with infinitesimal multiples of the unit operator are closed under the formation of commutators and therefore constitute the infinitesimal generators of a group, which we call the special canonical group.

4.1 DIFFERENTIAL OPERATORS

The transformations of this group can be usefully studied by their effect on the eigenvectors of the complete set of commuting Hermitian operators provided by the n_1 canonical variables q, or p, at a given time t. We first consider the interpretation of the transformation generated by G_q on the eigenvectors of the q-description, and similarly for the p-description, as indicated by (2.58),

$$\delta_q \langle q't| = i \langle q't| G_q$$
$$\delta_p \langle p't| = i \langle p't| G_p \quad . \qquad (4.4)$$

Now the vector $\langle q't| + \delta\langle q't|$ is an eigenvector of the operators $q - \delta q$ with the eigenvalues q'. But, since the δq are numerical multiples of the unit operator, the varied vector is described equivalently as an eigenvector of the operators q with the eigenvalues $q' + \delta q$. This shows that the spectra of the Hermitian operators q_k form a continuum, extending from $-\infty$ to ∞, and that the variation $\delta_q \langle q't|$ can be ascribed to a change of the eigenvalues q' by δq. A similar argument applies to the complementary variables p. Accordingly,

$$\delta_q \langle q't| = \sum_k \delta q_k \frac{\partial}{\partial q_k'} \langle q't|$$
$$\delta_p \langle p't| = \sum_k \delta p_k \frac{\partial}{\partial p_k'} \langle p't| \qquad (4.5)$$

and

$$\frac{1}{i} \frac{\partial}{\partial q_k'} \langle q't| = \langle q't| p_k(t)$$

$$i \frac{\partial}{\partial p_k'} \langle p't| = \langle p't| q_k(t) \quad . \tag{4.6}$$

The adjoint relations are

$$p_k(t) |q't\rangle = |q't\rangle \, i \frac{\partial^T}{\partial q_k'}$$

$$q_k(t) |p't\rangle = |p't\rangle \frac{1}{i} \frac{\partial^T}{\partial p_k'} \quad , \tag{4.7}$$

where the symbol ∂^T is used to indicate that the conventional sense of differentiation is reversed. For functions $F(q, p)$ that can be constructed algebraically from the variables p and arbitrary functions of the q, we establish by induction that

$$\langle q't| F(q, p) = F\left(q', -i \frac{\partial}{\partial q'}\right) \langle q't| \quad ,$$

and (4.8)

$$F(q, p) |q't\rangle = |q't\rangle F\left(q', i \frac{\partial^T}{\partial q'}\right) \quad ,$$

which supplies a differential operator realization

for the abstract operators. Similarly, for functions derived by algebraic construction from the q and arbitrary functions of the p ,

$$\langle p't|F(q\ ,\ p) = F\left(i\ \frac{\partial}{\partial p'}\ ,\ p'\right)\ \langle p't|$$

$$F(q\ ,\ p)\ |p't\rangle = |p't\rangle\ F\left(-i\ \frac{\partial^{T}}{\partial p'}\ ,\ p'\right)\ . \tag{4.9}$$

The significance of the transformation generated by G_p on the eigenvectors of the q - description, and conversely, is indicated by

$$\delta_p\langle q't| = i\langle q't|G_p = -\ i\langle q't|\ \sum_k \delta p_k\ q_k' \tag{4.10}$$

and

$$\delta_q|p't\rangle = -\ i\ G_q|p't\rangle = -i\ \sum_k p_k'\ \delta q_k\ |p't\rangle\ , \tag{4.11}$$

namely, these transformations multiplv the vectors by numerical phase factors. The transformations of which infinitesimal real multiples of the unit operator are the generators also multiply vectors by phase factors, but without distinction between the two descriptions. Hence, in response to the operations of the special canonical group, any eigenvector in the q or p descriptions with specified eigenvalues can be transformed into one with any other set of eigenvalues, and multiplied with an arbitrary phase factor.

4.2 SCHRÖDINGER EQUATIONS

The infinitesimal operator $G_t = -H\,\delta t$ generates the group of time translations, whereby a description at any one time is transformed into the analogous description at any other time. Thus, for the infinitesimal transformation from the q - description at time t to the q -description at time $t + \delta t$ we have

$$\delta_t\langle q't| = \delta t \frac{\partial}{\partial t} \langle q't| = i\langle q't|G_t \quad , \tag{4.12}$$

or, as an application of the differential operator realization (4.8), for a system described by variables of the first kind,

$$\begin{aligned} i \frac{\partial}{\partial t} \langle q't| &= \langle q't|\, H(q \;, p \;, t) \\ &= H\left(q' \;, -i \frac{\partial}{\partial q'} \;, t\right) \langle q't| \quad , \end{aligned} \tag{4.13}$$

if H is an algebraic function of the p variables. The adjoint statement can be written

$$|q't\rangle \frac{1}{i}\frac{\partial^T}{\partial t} = |q't\rangle H\left(q' , i\frac{\partial^T}{\partial q'} , t\right) . \qquad (4.14)$$

The arbitrary state symbolized by Ψ is represented by the complex conjugate pair of wave functions

$$\psi(q't) = \langle q't|\Psi , \qquad \phi(q't) = \Psi^{\dagger}|q't\rangle , \qquad (4.15)$$

and the variation of these functions in time is thus described by the (Schrödinger) differential equations

$$i\frac{\partial}{\partial t}\psi(q't) = H\left(q' , -i\frac{\partial}{\partial q'} , t\right)\psi(q't)$$
$$\phi(q't)\frac{1}{i}\frac{\partial^T}{\partial t} = \phi(q't) H\left(q' , i\frac{\partial^T}{\partial q'} , t\right) . \qquad (4.16)$$

4.3 THE q p TRANSFORMATION FUNCTION

The transformation function connecting the q and p representations at a common time can be constructed by integration of the differential equation

$$\begin{aligned}\delta \langle q't|p't\rangle &= i \langle q't|\left(G_q - G_p\right)|p't\rangle \\ &= i \sum_k \left(\delta q_k' \, p_k' + q_k' \, \delta p_k' \right) \langle q't|p't\rangle \\ &= \delta\left[i \sum_k q_k' p_k' \right] \langle q't|p't\rangle\end{aligned} \qquad (4.17)$$

in which we have made direct use of the operator properties contained in (4.10) and (4.11). Hence

$$\langle q't|p't\rangle = C \, e^{iq'p'} \qquad (4.18)$$

and then

$$\langle p't|q't\rangle = C^* \, e^{-ip'q'} \quad . \qquad (4.19)$$

The magnitude of the constant C is fixed by the composition property

$$\begin{aligned}\delta(q'-q'') &= \int \langle q't|p't\rangle \, dp' \, \langle p't|q''t\rangle \\ &= |C|^2 \int dp' \, e^{i \sum (q'-q'')_k \, p_k'} \quad ,\end{aligned} \qquad (4.20)$$

while the phase of C is an intrinsically arbitrary constant which can be altered freely by common phase factor transformations of the q states relative to the p states. Hence, with a conventional choice of phase, we have

$$C = (2\pi)^{-\frac{1}{2}n_1} \quad . \tag{4.21}$$

According to the structure of this transformation function, eigenvectors in the q and p representations are related by the reciprocal Fourier transformations

$$\begin{aligned} \langle p't| &= (2\pi)^{-n/2} \int e^{-ip'q'} dq' \langle q't| \\ \langle q't| &= (2\pi)^{-n/2} \int e^{iq'p'} dp' \langle p't| \quad . \end{aligned} \tag{4.22}$$

With the aid of these connections, the unit operator can be exhibited, not only in terms of the complete set of q, or p states,

$$1 = \int |q't\rangle dq' \langle q't| = \int |p't\rangle dp' \langle p't| \tag{4.23}$$

but also in mixed qp or pq representations,

$$\begin{aligned} 1 &= (2\pi)^{-n/2} \int |q't\rangle dq' e^{iq'p'} dp' \langle p't| \\ &= (2\pi)^{-n/2} \int |p't\rangle dp' e^{-ip'q'} dq' \langle q't| \quad . \end{aligned} \tag{4.24}$$

4.4 DIFFERENTIAL STATEMENTS OF COMPLETENESS

Alternative forms, involving differentiation rather than integration, can be given to these latter expressions of completeness, either by direct transformation, or from the following considerations. An arbitrary vector Ψ is constructed from the states $|q't\rangle$ and the wave function $\psi(q't)$ as (suppressing the reference to time)

$$\begin{aligned}\Psi &= \int |q'\rangle\, dq'\, \psi(q') = \Psi(q) \int |q'\rangle\, dq' \\ &= \psi(q)|p'=0\rangle (2\pi)^{n/2} \quad ,\end{aligned} \tag{4.25}$$

in which we have introduced the operator $\psi(q)$, and recognized the structure of the state $|p'\rangle$ with eigenvalues $p'=0$. Employing a similar procedure, we find

$$\Phi = \int \phi(p')\, dp' \langle p'| = (2\pi)^{n/2} \langle q'=0|\phi(p) \quad , \tag{4.26}$$

and thus the scalar product of two vectors can be evaluated as

$$\Phi_1\Psi_2 = (2\pi)^n \langle q'=0|\phi_1(p)\psi_2(q)|p'=0\rangle \quad . \tag{4.27}$$

The use of the differential operator realizations (4.8) or (4.9) now gives

$$\Phi_1\Psi_2 = (2\pi)^n \, \phi_1\left(-i \frac{\partial}{\partial q'}\right) \psi_2(q') \, \langle q'|p'=0\rangle \Big|_{q'=0}$$

$$= (2\pi)^{n/2} \, \phi_1\left(-i \frac{\partial}{\partial q'}\right) \psi_2(q') \Big|_{q'=0} \quad . \quad (4.28)$$

or

$$\Phi_1\Psi_2 = (2\pi)^n \langle q'=0|p'\rangle \, \phi_1(p') \, \psi_2\left(-i \frac{\partial^T}{\partial p'}\right) \Big|_{p'=0}$$

$$= (2\pi)^{n/2} \, \phi_1(p') \, \psi_2\left(-i \frac{\partial^T}{\partial p'}\right) \Big|_{p'=0} \quad . \quad (4.29)$$

On abstracting from the arbitrary vectors, these results become

$$1 = (2\pi)^{n/2} \left| -i \frac{\partial}{\partial q'} \right\rangle \left\langle q' \right|_{q'=0}$$
$$= (2\pi)^{n/2} \left| p' \right\rangle \left\langle -i \frac{\partial^T}{\partial p'} \right|_{p'=0} \quad . \quad (4.30)$$

and the adjoint operation produces the analogous properties

$$1 = (2\pi)^{n/2} \left| i \frac{\partial}{\partial p'} \right\rangle \left\langle p' \right|_{p'=0}$$

$$= (2\pi)^{n/2} \left| q' \right\rangle \left\langle i \frac{\partial^T}{\partial q'} \right|_{q'=0} \quad . \tag{4.31}$$

The practical utility of these forms depends, of course, upon the differentiation operations appearing in an algebraic manner. The basic examples of this type occur in the differential composition properties of the transformation functions $\langle q'|p'\rangle$ and $\langle p'|q'\rangle$,

$$\langle q'|p'\rangle = (2\pi)^{n/2} \left\langle q' \left| -i \frac{\partial}{\partial q} \right\rangle \langle q|p'\rangle \right|_{q=0} \tag{4.32}$$

and

$$\langle p'|q'\rangle = (2\pi)^{n/2} \langle p'|q\rangle \left\langle i \frac{\partial^T}{\partial q} \right| q' \Big\rangle \Big|_{q=0} \quad . \tag{4.33}$$

4.5 NON-HERMITIAN CANONICAL VARIABLES

The significance of the special canonical group will now be studied for the non-Hermitian canonical variables y , $iy^\dagger$, which are defined by

$$y_k = 2^{-\frac{1}{2}}(q_k + ip_k)$$
$$iy_k^\dagger = 2^{-\frac{1}{2}}(p_k + iq_k) \quad . \tag{4.34}$$

(In this discussion, we restrict the symbols q, p to represent Hermitian canonical variables). The formal theory provides the infinitesimal generators

$$G_y = iy^\dagger \, \delta y \quad , \quad G_y{}^\dagger = -i \, \delta y^\dagger \, y \tag{4.35}$$

and these operators obey the commutation relations

$$[G_y^{(1)} \, , \, G_y^{(2)}] = [G_{y^\dagger}^{(1)} \, , \, G_{y^\dagger}^{(2)}] = 0$$
$$[G_{y^\dagger}^{(1)} \, , \, G_y^{(2)}] = \delta^{(1)}y^\dagger \, \delta^{(2)}y \quad . \tag{4.36}$$

To follow the pattern set by the Hermitian canonical variables, we must first construct the eigenvectors of the two complete sets of commuting operators y and $y^\dagger$.

4.6 SOME TRANSFORMATION FUNCTIONS

The vectors $|y't\rangle$ are described in relation

to the states $\langle q't|$ by the transformation function $\langle q't|y't\rangle$, which obeys the differential equation

$$\delta \langle q't|y't\rangle = i\langle q't|(G_q - G_y)|y't\rangle \quad . \qquad (4.37)$$

Now the following forms of the generators,

$$iG_q = ip\,\delta q = \delta q\,(2^{\frac{1}{2}}y - q) \quad , \qquad (4.38)$$

and

$$-iG_y = y^{\dagger}\,\delta y = (2^{\frac{1}{2}}q - y)\,\delta y \quad , \qquad (4.39)$$

permit all operators to be replaced by their eigenvalues,

$$\begin{aligned} &\delta \langle q't|y't\rangle \\ &= [\,\delta q'\,(2^{\frac{1}{2}}y'-q') + (2^{\frac{1}{2}}q'-y')\,\delta y'\,]\,\langle q't|y't\rangle \\ &= \delta[-\tfrac{1}{2}q'^2 + 2^{\frac{1}{2}}q'y' - \tfrac{1}{2}y'^2]\,\langle q't|y't\rangle \quad , \end{aligned} \qquad (4.40)$$

which gives the desired transformation function,

$$\langle q't|y't\rangle = C\,\exp\left(-\tfrac{1}{2}q'^2+2^{\frac{1}{2}}q'y'-\tfrac{1}{2}y'^2\right) \quad . \qquad (4.41)$$

Since the adjoint of the right eigenvector equation

$$y|y'\rangle = y'|y'\rangle \tag{4.42}$$

is the left eigenvector equation for $y^\dagger$,

$$\langle y^{\dagger\prime}|y^\dagger = \langle y^{\dagger\prime}|y^{\dagger\prime} \quad , \quad y^{\dagger\prime} = y'^* \quad , \tag{4.43}$$

we conclude that the complex conjugate of (4.41) is

$$\langle y^{\dagger\prime}t|q't\rangle = C^* \exp\left(-\tfrac{1}{2}q'^2+2^{\frac{1}{2}}y^{\dagger\prime}q'-\tfrac{1}{2}y^{\dagger\prime 2}\right) . \tag{4.44}$$

We can now compute the transformation function

$$\langle y^{\dagger\prime}t|y''t\rangle = \int \langle y^{\dagger\prime}t|q't\rangle\, dq'\, \langle q't|y''t\rangle$$
$$= |C|^2\, \pi^{n/2}\, e^{y^{\dagger\prime}y''} , \tag{4.45}$$

which is the analogue of $\langle p't|q't\rangle$. In particular, the scalar product of a vector $|y't\rangle$ with its adjoint is given by

$$\langle y^{\dagger\prime}t|y't\rangle = |C|^2\, \pi^{n/2}\, e^{|y'|^2} \tag{4.46}$$

and this is a finite number. Hence the eigenvectors $|y't\rangle$ and $\langle y^{\dagger\prime}t|$ exist for all complex values of y' . It will be noted that the lengths of these vectors depend upon the eigenvalues, the eigenvector belonging to zero eigenvalues being uniquely distinguished as the one of minimum length. On normalizing the latter to unity, the constant C is determined in magnitude, and, with a conventional choice of phase, we have

$$\langle y^{\dagger\prime} t | y'' t\rangle = e^{y^{\dagger\prime} y''} \quad . \tag{4.47}$$

A similar discussion for the transformation function $\langle q' t | y^{\dagger\prime} t\rangle$ gives

$$\langle q' t | y^{\dagger\prime} t\rangle = C' \exp\left(+\tfrac{1}{2}q'^2 - 2^{\frac{1}{2}} q' y^{\dagger\prime} + \tfrac{1}{2} y^{\dagger\prime 2}\right) \quad , \tag{4.48}$$

and no vector $|y^{\dagger\prime} t\rangle$, or any linear combination of these vectors, possesses a finite length. This asymmetry between y and $y^{\dagger}$, in striking contrast with the situation for Hermitian canonical variables is evident from the commutation relation

$$y_k y_k^{\dagger} - y_k^{\dagger} y_k = 1 \quad , \tag{4.49}$$

for the substitution $y \to y^{\dagger}$, $y^{\dagger} \to -y$, while formally preserving the commutation relation, converts the non-negative operator $y^{\dagger} y$ into the non-positive operator $-yy^{\dagger}$. Thus, unlike $|y' t\rangle$

and $\langle y^{\dagger\prime} t|$, the vectors $|y^{\dagger\prime} t\rangle$ and $\langle y' t|$ do not exist. Yet we shall find it possible to define vectors which, in a limited sense, are right eigenvectors of $y^{\dagger}$ and left eigenvectors of y .

4.7 PHYSICAL INTERPRETATION

The physical interpretation of the states $|y' t\rangle$ requires some comment. Since these vectors are not normalized to unity, in general, we must express the expectation value for such states as

$$\langle F\rangle_{y'} = \langle y^{\dagger\prime}|F|y'\rangle / \langle y^{\dagger\prime}|y'\rangle \quad . \tag{4.50}$$

Now

$$\begin{aligned} \langle (y_k - y_k')\rangle_{y'} &= \langle (y_k - y_k')^2\rangle_{y'} \\ &= \langle (y_k^{\dagger} - y_k^{\dagger\prime})(y_k - y_k')\rangle_{y'} = 0 \quad , \end{aligned} \tag{4.51}$$

and, on writing

$$y' = 2^{-\frac{1}{2}}(q' + ip') \quad , \tag{4.52}$$

we infer that

$$\langle q_k\rangle_{y'} = q_k' \quad , \quad \langle p_k\rangle_{y'} = p_k' \tag{4.53}$$

and

$$\langle (q_k - q_k')^2 \rangle_{y'} = \langle (p_k - p_k')^2 \rangle_{y'} = \tfrac{1}{2} \quad . \qquad (4.54)$$

Thus, the states $|y't\rangle$ are such that neither the q nor the p variables have definite values but, rather, the distribution about the expectation values corresponds to an (optimum) compromise between the complementary aspects of the two incompatible sets of variables. We shall designate the normalized vectors of this description as

$$|q'p't\rangle = e^{-\frac{1}{2}|y'|^2} \, |y't\rangle \quad , \qquad (4.55)$$

in which notation the transformation function (4.47) reads

$$\langle q'p't|q''p''t\rangle$$
$$= e^{\frac{1}{2}i(q'p''-p'q'')} \, e^{-\frac{1}{4}(q'-q'')^2 - \frac{1}{4}(p'-p'')^2} \quad . \qquad (4.56)$$

The interpretation of the transformations produced by $G_{y^\dagger}$, and G_y, on the states $\langle y^{\dagger\prime}t|$, and $|y't\rangle$, respectively, is quite the same as for the Hermitian variables, which the construction of (4.41) implicitly assumes. Thus

$$<y^{\dagger\prime}t|y_k(t) = \frac{\partial}{\partial y_k^{\dagger\prime}} <y^{\dagger\prime}t|$$
$$y_k^{\dagger}(t)|y't> = |y't> \frac{\partial^T}{\partial y_k'} \quad , \tag{4.57}$$

and generally,

$$<y^{\dagger\prime}t|F(y^{\dagger}, y) = F\left(y^{\dagger\prime}, \frac{\partial}{\partial y^{\dagger\prime}}\right) <y^{\dagger\prime}t|$$
$$F(y^{\dagger}, y) |y't> = |y't> F\left(\frac{\partial^T}{\partial y'}, y'\right) \quad . \tag{4.58}$$

Accordingly, the wave functions $\psi(y^{\dagger\prime}t)$ and $\phi(y't)$ for a system described by non-Hermitian canonical variables of the first kind obey the Schrodinger equations

$$i \frac{\partial}{\partial t} \psi(y^{\dagger\prime}t) = H\left(y^{\dagger\prime}, \frac{\partial}{\partial y^{\dagger\prime}}, t\right) \psi(y^{\dagger\prime}t) \tag{4.59}$$

and

$$\phi(y't) \frac{1}{i} \frac{\partial^T}{\partial t} = \phi(y't) H\left(\frac{\partial^T}{\partial y'}, y', t\right) \quad . \tag{4.60}$$

The transformations generated by G_y on $<y^{\dagger\prime}t|$ and by $G_{y^{\dagger}}$ on $|y't>$ are given by

$$\delta_y \langle y^{\dagger'} t| = - \langle y^{\dagger'} t| \sum_k y_k^{\dagger'} \, \delta y_k$$
$$\delta_{y^\dagger} |y't\rangle = - \sum_k \delta y_k^\dagger \, y_k' \, |y't\rangle \quad . \tag{4.61}$$

These changes are infinitesimal multiples of the original vectors, but unlike the situation with Hermitian variables, the multiplicative factors are not necessarily imaginary numbers. The same comment applies to the infinitesimal multiples of the unit operator produced by commuting two generators, as in (4.36). Hence the special canonical group for non-Hermitian variables constitutes the totality of transformations that change an eigenvector with specified complex eigenvalues into one with any other set of eigenvalues, and which is multiplied by arbitrary complex numerical factors.

4.8 COMPOSITION BY CONTOUR INTEGRATION

The transformation function $\langle y^{\dagger'} t|y''t\rangle$ differs in form from $\langle p't|q't\rangle$ only in the absence of the powers of 2π, and therefore obeys the differential composition property

$$\left\langle y^{\dagger\prime} t \left| \frac{\partial}{\partial y^{\dagger}} t \right\rangle \langle y^{\dagger} t | y'' t \rangle \right|_{y^{\dagger}=0} = \langle y^{\dagger\prime} t | y'' t \rangle \;. \quad (4.62)$$

We infer that

$$1 = \left| \frac{\partial}{\partial y^{\dagger\prime}} t \right\rangle \left\langle y^{\dagger\prime} t \right|_{y^{\dagger\prime}=0} = \langle y' t \rangle \left\langle \frac{\partial^T}{\partial y'} t \right|_{y'=0} \;, \quad (4.63)$$

or, that scalar products of arbitrary vectors can be computed as

$$\begin{aligned} \Phi_1 \Psi_2 &= \phi_1\left(\frac{\partial}{\partial y^{\dagger\prime}} t \right) \psi_2(y^{\dagger\prime} t) \Big|_{y^{\dagger\prime} = 0} \\ &= \phi_1(y' t) \; \psi_2\left(\frac{\partial^T}{\partial y'} t \right) \Big|_{y'=0} \;. \end{aligned} \quad (4.64)$$

Now one can give the evaluation at zero of these complex eigenvalues a contour integral form (in this discussion we place $n=1$, for simplicity; the extension to arbitrary n is immediate). Thus,

$$\begin{aligned}\Phi_1\Psi_2 &= \frac{1}{2\pi i}\oint \frac{dy^{\dagger'}}{y^{\dagger'}}\,\phi_1\Big(\frac{\partial}{\partial y^{\dagger'}}\,t\Big)\,\psi_2\big(y^{\dagger'}t\big) \\ &= \frac{1}{2\pi i}\oint \phi_1(y't)\,\psi_2\Big(\frac{\partial^T}{\partial y'}\,t\Big)\,\frac{dy'}{y'} \quad ,\end{aligned} \tag{4.65}$$

where the integration paths encircle the origin in the positive sense and enclose only the singularity at the origin. Such paths can be drawn since the existence of the limits in (4.64) requires that wave functions of the type $\psi(y^{\dagger'}t)$ and $\phi(y't)$ be regular functions of the corresponding complex variable in a neighborhood of the origin. Alternative forms obtained by partial integration are

$$\begin{aligned}\Phi_1\Psi_2 &= \frac{1}{2\pi i}\oint dy^{\dagger'}\;\phi_1(y^{\dagger'}t)\;\psi_2(y^{\dagger'}t) \\ &= \frac{1}{2\pi i}\oint dy'\;\phi_1(y't)\;\psi_2(y't) \quad ,\end{aligned} \tag{4.66}$$

where

$$\begin{aligned}\frac{1}{y^{\dagger'}}\;\phi\Big(-\frac{\partial^T}{\partial y^{\dagger'}}\,t\Big) &= \phi\big(y^{\dagger'}t\big) \quad , \\ \psi\Big(-\frac{\partial}{\partial y'}\,t\Big)\,\frac{1}{y'} &= \psi(y't) \quad .\end{aligned} \tag{4.67}$$

These functions are regular in a neighborhood of the point at infinity. The integrations in

(4.66) are to be extended through a common domain of regularity of the two factors, with the enclosed region containing all singularities of $\phi(y^{\dagger\prime}t)$, or $\psi(y't)$, but no singularity of $\psi(y^{\dagger\prime}t)$, or $\phi(y't)$.

If the wave functions $\psi(y't)$ and $\phi(y^{\dagger\prime}t)$ are related to the vector bases $\langle y't|$ and $|y^{\dagger\prime}t\rangle$, given symbolically by

$$\langle y't| = \left\langle -\frac{\partial}{\partial y'}\, t\right| \frac{1}{y'} \tag{4.68}$$

and

$$|y^{\dagger\prime}\rangle = \frac{1}{y^{\dagger\prime}} \left| -\frac{\partial^T}{\partial y^{\dagger\prime}}\, t\right\rangle , \tag{4.69}$$

the scalar product evaluations can be presented as the completeness properties

$$\begin{aligned} 1 &= \frac{1}{2\pi i} \oint dy' \, |y't\rangle\langle y't| \\ &= \frac{1}{2\pi i} \oint dy^{\dagger\prime} \, |y^{\dagger\prime}t\rangle\langle y^{\dagger\prime}t| \quad . \end{aligned} \tag{4.70}$$

It is implied by these statements that

$$\frac{1}{2\pi i} \oint dy' \, \phi(y't) \, \langle y't|y''t\rangle = \phi(y''t) \tag{4.71}$$

and

$$\frac{1}{2\pi i} \oint dy'' \, \langle y't|y''t\rangle \, \psi(y''t) = \psi(y't) \quad , \qquad (4.72)$$

where, according to the definition (4.68), and the transformation function (4.47),

$$\begin{aligned} \langle y't|y''t\rangle &= \left\langle -\frac{\partial}{\partial y'} \, t \middle| y''t \right\rangle \frac{1}{y'} = e^{-y'' \frac{\partial}{\partial y'}} \frac{1}{y'} \\ &= \frac{1}{y' - y''} \quad , \qquad |y'| > |y''| \quad . \end{aligned} \qquad (4.73)$$

The indicated regularity domain corresponds to that of wave functions of the types $\psi(y')$ and $\phi(y'')$. Indeed,

$$\frac{1}{2\pi i} \oint dy' \, \frac{\phi(y't)}{y' - y''} = \phi(y''t) \qquad (4.74)$$

if the contour encloses y'' but contains no singularity of $\phi(y't)$, while

$$\frac{1}{2\pi i} \oint dy'' \, \frac{\psi(y''t)}{y' - y''} = \psi(y't) \qquad (4.75)$$

if the integration path encloses a region continuing all the singularities of $\psi(y''t)$ but not the point y' . In the latter situation, the integral is evaluated with the aid of a circle about the point at infinity where the function $\psi(y''t)$ vanishes. A similar discussion applies to the properties of the transformation function

$$\langle y^{\dagger\prime}t|y^{\dagger\prime\prime}t\rangle = \frac{1}{y^{\dagger\prime\prime} - y^{\dagger\prime}} \quad , \qquad |y^{\dagger\prime\prime}| > |y^{\dagger\prime}| \quad . \tag{4.76}$$

The symbolic construction of $\langle y't|$ is made explicit on writing

$$\frac{1}{y'} = \int_0^\infty dy^{\dagger\prime\prime}\, e^{-y^{\dagger\prime\prime}y'} \quad , \tag{4.77}$$

where the integration path is extended to infinity along any path that implies convergence of this, and subsequent integrals. Thus

$$\langle y't| = \int_0^\infty dy^{\dagger\prime\prime}\, \langle y^{\dagger\prime\prime}t|e^{-y^{\dagger\prime\prime}y'} \tag{4.78}$$

and

$$|y^{\dagger\prime}t\rangle = \int_0^\infty dy''\, e^{-y^{\dagger\prime}y''}\, |y''t\rangle \quad . \tag{4.79}$$

With these forms we can examine to what extent $\langle y't|$ and $|y^{\dagger\prime}t\rangle$ are left eigenvector of $y(t)$, and right eigenvector of $y^{\dagger}(t)$, respectively.

Now

$$\langle y't|(y-y') = \int_0^{\infty} dy^{\dagger''}\, e^{-y^{\dagger''}y'} \left(\frac{\partial}{\partial y^{\dagger''}} - y'\right) \langle y^{\dagger''}t|$$

$$= \int_0^{\infty} dy^{\dagger''} \frac{\partial}{\partial y^{\dagger''}} \left(e^{-y^{\dagger''}y'} \langle y^{\dagger''}t|\right)$$

$$= - \langle y^{\dagger''}=0\;,\; t| \quad , \tag{4.80}$$

or, as we can recognize from the symbolic form (4.68)

$$\langle y't|y = \langle y't|y' - \frac{1}{2\pi i}\oint dy'' \langle y''t| \quad , \tag{4.81}$$

provided the contour encircles all the singularities of the integrand. The adjoint statement is

$$y^{\dagger}|y^{\dagger'}t\rangle = y^{\dagger'}|y^{\dagger'}t\rangle - \frac{1}{2\pi i}\oint dy^{\dagger''}\, |y^{\dagger''}t\rangle \quad . \tag{4.82}$$

Hence, as our discussion of the asymmetry between y and $y^{\dagger}$ would lead us to anticipate, $\langle y't|$ and $|y^{\dagger'}t\rangle$ are not true eigenvectors[1], although

---

[1] The use of complex eigenvalues has been developed in a more formal way by P.A.M. Dirac, Comm. Dub. Inst. for Adv. Studies, Ser. A. No. 1 (1943), without initial recognition of the asymmetry be-

the failure of this property appears in a very simple form. The consistency of the theory, in which the action of y on its left eigenvectors contains the additional contour integral term of (4.81), can be verified from the alternative evaluations of $\langle y'|y|y''\rangle$ where, acting on the right eigenvector,

$$\langle y'|y|y''\rangle = \langle y'|y''\rangle y'' = \frac{y''}{y'-y''} , \tag{4.83}$$

whereas the properties of the left eigenvectors give

$$\begin{aligned} \langle y'|y|y''\rangle &= y'\langle y'|y''\rangle - \frac{1}{2\pi i}\oint dy \, \langle y|y''\rangle \\ &= \frac{y'}{y'-y''} - 1 = \frac{y''}{y'-y''} . \end{aligned} \tag{4.84}$$

4.9 MEASUREMENTS OF OPTIMUM COMPATIBILITY

According to the eigenvector constructions

---

tween left and right eigenvectors of the non-Hermitian variables. With this procedure, the alternative evaluations of $\langle y'|y|y''\rangle$ lead to results differing by unity and, generally, one is forced to assume that two $\psi(y')$ functions differing by an arbitrary non-negative power series in y' describe the same state.

(4.78) and (4.79), the unit vector can be presented as

$$1 = \int \frac{dy'\, dy^{\dagger''}}{2\pi i} \, |y't\rangle \, e^{-y'y^{\dagger''}} \, \langle y^{\dagger''}t| \quad , \tag{4.85}$$

which is the analogue of (4.24). Here the complex variables y' and $y^{\dagger''}$ are to be integrated along orthogonal paths. If we write

$$y' = 2^{-\frac{1}{2}}(q'+ip') \quad , \quad y^{\dagger''} = 2^{-\frac{1}{2}}(q'-ip') \quad , \tag{4.86}$$

the paths can be so deformed that q' and p' are integrated independently from $-\infty$ to ∞ through the domain of real values. This produces the form

$$1 = \int \frac{dq'\, dp'}{2\pi} \, |y't\rangle \, e^{-|y'|^2} \, \langle y^{\dagger'}t| \tag{4.87}$$

or, in the notation of (4.55),

$$1 = \int |q'p't\rangle \, \frac{dq'\, dp'}{2\pi} \, \langle q'p't| \quad . \tag{4.88}$$

The transformation function composition property implied by this expression of completeness,

$$\langle q'p't|q''p''t\rangle$$

$$= \int \langle q'p't|qpt\rangle \frac{dq\, dp}{2\pi} \langle qpt|q''p''t\rangle \, , \qquad , \qquad (4.89)$$

can be verified by direct integration. Although the vectors $|q'p't\rangle$ are complete, they are certainly not linearly independent or the transformation function (4.56) would be a delta function. Instead, it appears that displacing the eigenvalues $q'p'$ from $q''p''$, by amounts of the order of unity, does not produce an essentially different state. Changes in eigenvalues that are considerably in excess of unity do result in new states, however, since the value of the transformation function becomes very small. In a rough sense, there is one state associated with each eigenvalue range $(\Delta q' \Delta p')/2\pi = 1$. It is not necessary to construct the linearly independent vectors that describe distinct states if we are interested only in the comparison with measurements on a classical level (as the final stage of every measurement must be) for then we are concerned with the probability that the system be encountered in one of a large number of states, corresponding to an eigenvalue range $(\Delta q' \Delta p')/2\pi >> 1$. Within this context, it can be asserted that

$$dp(q'p't\ ,\ \Psi) = \frac{dq'\ dp'}{2\pi}\,|\psi(q'p't)|^2 \qquad (4.90)$$

is the probability that q and p measurements, performed with optimum compatibility on the state ψ at the time t, will yield values q' and p' lying in the intervals dq' and dp', respectively.

II VARIABLES OF THE SECOND KIND

4.10 ROTATION GROUP

Now we turn to the variables of the second kind. The requirement that the variations $\delta\zeta$ anticommute with each variable of this type is expressed most simply with the aid of the canonical Hermitian variables ξ_κ. The enumeration of the $2^\nu = 4^{n^2}$ distinct operators of the algebra shows that there is only one operator with the property of anticommuting with every ξ. This is the product

$$2^{\frac{1}{2}}\,\xi_{2n+1} = (-2i)^n \xi_1 \xi_2\ , \ldots ,\ \xi_{2n}\ , \qquad (4.91)$$

which is written as a Hermitian operator with unit square. Thus the operator properties of the ξ variables contained in (3.110) also apply to ξ_{2n+1}. We see that the variations $\delta\xi_\kappa$ must be infinitesimal real multiples of the single operator ξ_{2n+1}, say

$$\tfrac{1}{2}\,\delta\xi_\kappa = -\,\delta\omega_\kappa\,\xi_{2n+1}\,, \tag{4.92}$$

so that the infinitesimal generator

$$G_\xi = \frac{i}{2}\,\xi\,\delta\xi \tag{4.93}$$

acquires the form

$$G_\xi = \sum_{\kappa=1}^{2n} \delta\omega_\kappa\,\frac{1}{i}\,\xi_\kappa\xi_{2n+1}\,. \tag{4.94}$$

On forming the commutator of two such generators we obtain

$$\frac{1}{i}\,[G_\xi^{(1)}\,,\,G_\xi^{(2)}] = \sum_{\kappa\lambda} \delta^{(1)}\omega_\kappa\,\delta^{(2)}\omega_\lambda\,\xi_{\kappa\lambda}\,, \tag{4.95}$$

where the

$$\xi_{\kappa\lambda} = \frac{1}{2i}\,[\xi_\kappa\,,\,\xi_\lambda] \tag{4.96}$$

comprise a set of $\frac{1}{2}\nu(\nu-1) = n(2n-1)$ Hermitian operators obeying

$$\xi_{k\lambda}{}^2 = \tfrac{1}{4} \quad . \tag{4.97}$$

Thus the generators G_ξ and their commutators can be constructed from the operator basis provided by the $n(2n+1)$ Hermitian operators $\xi_{\kappa\lambda}$, where κ and λ range from 1 to $2n+1$. This basis is complete according to the commutation property

$$\begin{aligned} &\frac{1}{i}\,[\xi_{\kappa\lambda}\ ,\ \xi_{\mu\nu}] \\ &= \delta_{\lambda\nu}\,\xi_{\kappa\mu} - \delta_{\kappa\nu}\,\xi_{\lambda\mu} + \delta_{\kappa\mu}\,\xi_{\lambda\nu} - \delta_{\lambda\mu}\,\xi_{\kappa\nu} \quad , \end{aligned} \tag{4.98}$$

and the totality of these transformations form a group, which possesses the structure of the (proper) rotation group in $2n+1$ dimensions [compare (2.80)] . But this is not what we shall call the special canonical group for variables of the second kind.

4.11 EXTERNAL ALGEBRA

We have been discussing a group of inner automorphisms - transformations, constructed from the elements of the algebra, that maintain all algebraic relations and Hermitian properties. It is characteristic of the structure of the algebra that the only operators anticommutative with every ξ_κ , $\kappa = 1,\ldots,2n$, are numerical multiples of ξ_{2n+1} , and two variations formed in this way are commutative. Hence the generator of one variation does not commute with a second such variation and two variations are not independent. To obtain independent variations, the operators $\delta^{(1)}\xi$ and $\delta^{(2)}\xi$ must anticommute, and this is impossible for inner automorphisms. But the algebraically desirable introduction of independent variations can be achieved - at the expense of Hermitian properties - by considering outer automorphisms, constructed with the aid of a suitably defined external algebra.

Let ε_κ be a set of 2n completely anticommutative operators,

$$\{\varepsilon_\mu , \varepsilon_\lambda\} = 0 , \tag{4.99}$$

that commute with the elements of the physical algebra. That anticommutativity property includes

$$\varepsilon_\kappa^{\,2} = 0 , \tag{4.100}$$

and these operators cannot be Hermitian, nor is the adjoint of any ε - operator included in the set.

We see that the $2n$ operator products $\varepsilon_\kappa \xi_{2n+1}$ are completely anticommutative among themselves, and also anticommute with every ξ_κ, $\kappa = 1,\ldots,2n$. Hence variations $\delta\xi_\kappa$ defined as numerical multiples of $\varepsilon_\kappa \xi_{2n+1}$ will obey

$$\{ \delta^{(1)}\xi_\kappa , \delta^{(2)}\xi_\lambda \} = 0 \tag{4.101}$$

and

$$[\delta^{(1)}\xi_\kappa , G_\xi^{(2)}] = 0 . \tag{4.102}$$

It is the latter property that unifies the generators for the variables of the first and second kind, and permits the commutator of two generators, $G_x^{(1)}$ and $G_x^{(2)}$, to be evaluated generally as

$$\frac{1}{i} [G_x^{(1)} , G_x^{(2)}] = \frac{\partial_r G_x^{(1)}}{\partial x} \tfrac{1}{2} \delta^{(2)}x = - \tfrac{1}{2}\delta^{(1)}x \frac{\partial_\ell G_x^{(2)}}{\partial x}$$

$$= - \tfrac{1}{4} \delta^{(1)}x \; A \; \delta^{(2)}x . \tag{4.103}$$

The canonical form is (4.3) which, as written, applies to either type of variable. For the variables of the second kind, $\delta^{(1)}x \, A \, \delta^{(2)}x$ is linearly related to the products $\varepsilon_\kappa \varepsilon_\lambda$, and these combinations commute with all operators of the physical algebra and of the external algebra. Hence, with either kind of variable the generators of independent variations together with their commutators form the infinitesimal elements of a group, which is the special canonical group.

4.12 EIGENVECTORS AND EIGENVALUES

The existence of this group for the variables of the second kind enables one to define eigenvectors of the complete set of anticommuting operators q or $p = iq^{\dagger}$. Let us observe first that only the algebraic properties of the operators and variations are involved in the infinitesimal transformation equations

$$\begin{aligned}(1-iG_q)q_\kappa(1+iG_q) &= q_\kappa - \frac{1}{i}[q_\kappa , G_q] \\ &= q_\kappa - \delta q_\kappa\end{aligned} \tag{4.104}$$

and

$$(1-iG_{q^\dagger})\ q_\kappa^\dagger\ (1+iG_{q^\dagger}) = q_\kappa^\dagger - \delta q_\kappa^\dagger\ . \tag{4.105}$$

Furthermore, if q_κ' and $q_\kappa^{\dagger\prime}$ are quantities formed in the same way as the independent variations, anticommuting among themselves and with the operators q and $q^\dagger$, we can assert that

$$\begin{aligned} (1-iG_q)(q_\kappa - q'_\kappa)(1+iG_q) &= q_\kappa - (q'_\kappa + \delta q_\kappa) \\ (1-iG_{q^\dagger})(q^\dagger_\kappa - q^{\dagger\prime}_\kappa)(1+iG_{q^\dagger}) &= q^\dagger_\kappa - (q^{\dagger\prime}_\kappa + \delta q^\dagger_\kappa) \quad . \end{aligned} \tag{4.106}$$

Hence if vectors exist, obeying

$$(q_\kappa - q'_\kappa)|q't\rangle = 0 \tag{4.107}$$

and

$$\langle q^{\dagger\prime}t|(q^\dagger_\kappa - q^{\dagger\prime}_\kappa) = 0 \quad , \tag{4.108}$$

so also do the vectors

$$\begin{aligned} (1-iG_q)|q't\rangle &= |q't\rangle + \delta|q't\rangle \\ \langle q^{\dagger\prime}t|(1+iG_{q^\dagger}) &= \langle q^{\dagger\prime}t| + \delta\langle q^{\dagger\prime}t| \end{aligned} \tag{4.109}$$

exist, and these are similarly related to the eigenvalues $q' + \delta q$ and $q^{\dagger\prime} + \delta q^\dagger$, respectively. We shall see that that the eigenvectors associated with zero eigenvalues certainly exist, which implies that eigenvectors of the type $|q't\rangle$ and $\langle q^{\dagger\prime}t|$

can be constructed from the null eigenvalue states by the operations of the special canonical group. It should be remarked that these right and left eigenvectors are not in adjoint-relationship, since there is no such connection between q' and $q^{\dagger\prime}$.

A vector obeying the equations

$$q_\kappa|0t\rangle = 0 \qquad \kappa = 1,\ldots,n \tag{4.110}$$

is an eigenvector of the Hermitian operators $q_\kappa^\dagger q_\kappa$, with zero eigenvalues.

$$q_\kappa^\dagger q_\kappa|0t\rangle = 0 \qquad \kappa = 1,\ldots n \quad . \tag{4.111}$$

The converse is also true, since

$$\langle 0t|q_\kappa^\dagger q_\kappa|0t\rangle = 0 \tag{4.112}$$

implies (4.110). The operators $q_\kappa^\dagger q_\kappa$ are commutative and indeed constitute a complete set of commuting Hermitan operators for the variables of the second kind. (These statements apply equally to the non-Hermitian variables of the first kind.) And, from the algebraic properties of the canonical variables, we see that

$$q_\kappa^\dagger q_\kappa (1 - q_\kappa^\dagger q_\kappa) = q_\kappa^\dagger \;\; q_\kappa^2 \;\; q_\kappa^\dagger$$
$$= 0 \quad ; \tag{4.113}$$

the spectrum of each operator $q_\kappa^\dagger q_\kappa$ contains only the values 0 and 1 . Hence there is a state for which all operators $q_\kappa^\dagger q_\kappa$ possess the value zero and the state is also described by a right eigenvector of the non-Hermitian operators q , or the adjoint left eigenvector of the $q^\dagger$, belonging to the set of zero eigenvalues. We should observe here that, unlike the situation with non-Hermitian variables of the first kind, there is complete symmetry between the operators q and $q^\dagger$. In particular, the operators $q_\kappa q_\kappa^\dagger$ possess zero eigenvalues, which are equivalent to the unit eigenvalues of $q_\kappa^\dagger q_\kappa$, and the zero eigenvalue, right eigenvector of the $q^\dagger$ and left eigenvector of the q also exist. The possibility of defining the eigenvectors $|q^{\dagger\prime}t\rangle$ and $\langle q't|$ is then indicated by the differential equations

$$(1-iG_{q^\dagger})|q^{\dagger\prime}t\rangle = |q^{\dagger\prime}t\rangle + \delta|q^{\dagger\prime}t\rangle$$
$$\langle q't|(1+iG_q) = \langle q't| + \delta\langle q't| \quad , \tag{4.114}$$

which possess interpretations analogous to those of (4.109).

III UNIFICATION OF THE VARIABLES

4.13 CONSTRUCTIVE USE OF THE SPECIAL CANONICAL GROUP

In the progression from Hermitian canonical variables of the first kind to the non-Hermitian canonical variables of the second kind, the importance of the special canonical group has increased to the point where one uses it explicitly to define the eigenvectors of the canonical variables, rather than merely investigating the effect of the transformation group on independently constructed vectors. The former approach is universally applicable for, with all type of variables, the zero eigenvalue, right eigenvector of the canonical variables q can be constructed and the general eigenvector defined by a finite operation of the group,

$$|q'\rangle = e^{-ipq'}|q'=0\rangle \quad . \tag{4.115}$$

The analogous general construction of left eigenvectors is

$$\langle p'| = \langle p'=0|e^{-ip'q} \quad . \tag{4.116}$$

The Hermitian variables of the first kind and the non-Hermitian variables of the second kind permit, in addition, the construction of the vectors

$$\begin{aligned} |p'\rangle &= e^{ip'q}|p'=0\rangle \\ \langle q'| &= \langle q'=0|e^{ipq'} \quad , \end{aligned} \tag{4.117}$$

and the exceptional situation of the Hermitian canonical variables of the first kind stems from the unitary nature of the operator group for those variables, which deprives the zero eigenvalues of any distinguished position. The significance of the operations of the special canonical group on the eigenvectors of the canonical variables is indicated generally by

$$e^{-ipq''}|q'\rangle = |q'+q''\rangle \tag{4.118}$$

and

$$e^{-ip'q}|q'\rangle = e^{-ip'q'}|q'\rangle \quad ; \qquad (4.119)$$

they are the totality of transformations that alter eigenvalues and multiply eigenvectors by commutative factors.

The eigenvalues in the equations

$$q_\kappa|q'\rangle = q_\kappa'|q'\rangle \qquad (4.120)$$

and

$$\langle q'|q_\kappa = \langle q'|q_\kappa' \qquad (4.121)$$

are formed by multiplying an element of the external algebra with the member of the physical algebra that anticommutes with every ξ_κ , or equivalently, with the totality of operators q_κ and $q_\kappa{}^\dagger$. The translation of (4.91) into the language of the canonical variables presents the anticommutative operators as

$$2^{\frac{1}{2}}\xi_{2n+1} = \prod_{\kappa=1}^{n} (q_\kappa^\dagger q_\kappa - q_\kappa q_\kappa^\dagger) \quad , \qquad (4.122)$$

which we shall designate as ρ . This product, being a function of the commuting operators $q_\kappa^\dagger q_\kappa$, possesses definite values in the null eigenvalue states,

$$\left(\rho - (-1)^n\right)|q'=0> = 0 \quad , \quad <q'=0|(\rho-1) = 0 \ , \tag{4.123}$$

and the adjoint statements are

$$<q^{\dagger\prime}=0|\left(\rho - (-1)^n\right) = 0 \quad , \quad (\rho-1)|q^{\dagger\prime}=0> = 0 \ . \tag{4.124}$$

Accordingly, in considering

$$q_\kappa'|q'> = q_\kappa' e^{-ipq'}|q'=0> = e^{-ipq'}q_\kappa'|q'=0> \quad , \tag{4.125}$$

the factor ρ that occurs in q_κ' can be replaced by the number $(-1)^n$ and the result expressed by

$$q|q'> = q'|q'> = |q'>q' \tag{4.126}$$

where the final q' is entirely an element of the external algebra. In a similar way

$$<q'|q' = <q'=0|e^{ipq'}q' = <q'=0|q'e^{ipq'} \tag{4.127}$$

and

$$<q'|q = <q'|q' = q'<q'| \quad , \tag{4.128}$$

although the element of the external algebra that appears in the latter equation lacks the numerical factor $(-1)^n$ and therefore differs in sign from that of (4.126) if n is odd. Hence it is only for even n that complete symmetry between left and right eigenvectors exists, which invites the aesthetic judgment that no system described by an odd number (of pairs) of dynamical variables of the second kind exists in nature.

4.14 TRANSFORMATION FUNCTIONS

In constructing transformation functions, one must eliminate explicit reference to the operators of the physical system and express the transformation function in terms of the eigenvalues which, for variables of the second kind, are the elements of the external algebra. The replacement of eigenvalues that anticommute with the dynamical variables by purely external quantities is accomplished generally by equations of the type

$$q''|q'\rangle = |q'\rangle q'' \quad , \quad p'|q'\rangle = |q'\rangle p' \quad , \tag{4.129}$$

and occurs automatically for products of eigenvalues.

The transformation function $\langle p'|q'\rangle$ is characterized for all variables by the differential expression

$$\begin{aligned}\delta \langle p'|q'\rangle &= i\langle p'|\left(G_p - G_q\right)|q'\rangle \\ &= i\langle p'|(-\delta p'\, q' - p'\, \delta q')|q'\rangle \\ &= \delta[-ip'q']\,\langle p'|q'\rangle \quad ,\end{aligned} \qquad (4.130)$$

and differs in its integral form only by the numerical factors that express the normalization conventions for the particular type of variable. To achieve the universal form

$$\langle p'|q'\rangle = e^{-ip'q'} \qquad (4.131)$$

we must remove the factor $(2\pi)^{-n/2}$ that appears for Hermitian variables of the first kind. This will be done if all integrations are performed with the differentials

$$d[q'] = \frac{dq'}{(2\pi)^{n/2}} \qquad d[p'] = \frac{dp'}{(2\pi)^{n/2}} \qquad (4.132)$$

and delta functions correspondingly redefined:

$$\delta[q'] = (2\pi)^{n/2}\,\delta(q) \;, \quad \int d[q']\,\delta[q'] = 1 \;. \qquad (4.133)$$

With variables of the second kind, the transformation function can be expressed more specifically as

$$\langle q^{\dagger\prime}|q'\rangle = e^{q^{\dagger\prime}q'} = \prod_{\kappa} e^{q_{\kappa}^{\dagger\prime}q_{\kappa}'} = \prod_{\kappa} (1 + q_{\kappa}^{\dagger\prime}q_{\kappa}') \quad , \tag{4.134}$$

since the square of any eigenvalue vanishes. The $\langle p'|q'\rangle$ transformation function possesses the general differential composition property

$$\begin{aligned} \langle p'|q'\rangle &= \left\langle p'\right| i\,\frac{\partial_{\ell}}{\partial p}\left.\right\rangle \langle p|q'\rangle \Big|_{p=0} \\ &= \langle p'|q'\rangle \left\langle i\,\frac{\partial_r^T}{\partial q}\right|\left.q'\right\rangle \Big|_{q=0} \end{aligned} \tag{4.135}$$

which is expressed symbolically by

$$1 = \left|i\,\frac{\partial_{\ell}}{\partial p'}\right\rangle \left\langle p'\right| _{p'=0} = \left|q'\right\rangle \left\langle i\,\frac{\partial_r^T}{\partial q'}\right| _{q'=0} \quad . \tag{4.136}$$

Thus, for variables of the second kind, the scalar product of two vectors can be computed from the representative wave functions

$$\psi(q^{\dagger\prime}t) = \langle q^{\dagger\prime}t|\Psi \quad , \quad \phi(q't) = \Phi|q't\rangle \tag{4.137}$$

by

$$\Phi_1\Psi_2 = \phi_1\left(\frac{\partial_\ell}{\partial q^{\dagger'}}\, t\right)\, \psi_2(q^{\dagger'}t)\Big|_{q^{\dagger'}=0}$$

$$= \phi_1(q't)\, \psi_2\left(\frac{\partial^{\mathbf{T}}_r}{\partial q'}\, t\right)\Big|_{q'=0} \quad . \tag{4.138}$$

If non-Hermitian variables of the first kind are excepted, one can construct the transformation function

$$\langle q'|p'\rangle = e^{ip'q'} \tag{4.139}$$

which, for variables of the second kind, becomes

$$\langle q'|q^{\dagger'}\rangle = e^{-q^{\dagger'}q'} = e^{q'\,q^{\dagger'}} = \prod \left(1 + q'_K\, q^{\dagger'}_K\right) \quad . \tag{4.140}$$

We also have as the analogues of (4.137) and (4.138), the wave functions

$$\psi(q't) = \langle q't|\Psi \quad , \quad \phi(q^{\dagger'}t) = \Phi|q^{\dagger'}t\rangle \tag{4.141}$$

and the scalar product evaluation

$$\Phi_1\Psi_2 = \phi_1\left(\frac{\partial_\ell}{\partial q'}\, t\right) \psi_2(q't) \Bigg|_{q'=0}$$
$$= \phi_1(q^{\dagger\prime} t)\, \psi_2\left(\frac{\partial_r^T}{\partial q^{\dagger\prime}}\, t\right) \Bigg|_{q^{\dagger\prime}=0} \quad . \qquad (4.142)$$

With the same exception, the transformation function $\langle q'|q''\rangle$ and $\langle p'|p''\rangle$ are meaningful without qualification. The differential equation

$$\delta \langle q'|q''\rangle = i\langle q'|(p\, \delta q' - p\, \delta q'')|q''\rangle$$
$$= i\langle q'|p|q''\rangle\, \delta(q'-q'') \quad , \qquad (4.143)$$

combined with

$$0 = \langle q'|(q-q)|q''\rangle = \langle q'|q''\rangle(q'-q'') \qquad (4.144)$$

indicates that $\langle q'|q''\rangle$ is a function of the eigenvalue differences that vanishes on multiplication with any of its variables. (For non-hermitian variables of the first kind, the latter equation would read

$$0 = \langle y'|y''\rangle(y'-y'') - \frac{1}{2\pi i}\oint dy\, \langle y|y''\rangle \ , \qquad (4.145)$$

which is solved by $\langle y'|y''\rangle = (y'-y'')^{-1}$.) With Hermitian variables of the first kind these properties define the delta function,

$$\langle q'|q''\rangle = \delta[q'-q''] \quad , \tag{4.146}$$

and we shall retain this notation for the corresponding function referring to the second class of variable,

$$\langle q'|q''\rangle = \prod (q''_\kappa - q'_\kappa) = \delta[q'-q''] \quad , \tag{4.147}$$

in which the product of n anticommuting factors is arranged in some standard order, say $1,\ldots,n$, as read from left to right. For the similar transformation function referring to the variables $q^\dagger$ we write

$$\langle q^{\dagger\prime}|q^{\dagger\prime\prime}\rangle = \prod^{T} (q^{\dagger\prime}_\kappa - q^{\dagger\prime\prime}_\kappa) = \delta[q^{\dagger\prime} - q^{\dagger\prime\prime}] \quad , \tag{4.148}$$

with the reversed sense of multiplication. The consistency of these definitions follows from the composition property

$$\left\langle q^{\dagger\prime}\right| \frac{\partial_\ell}{\partial q} \Big\rangle \langle q|q'\rangle \Big|_{q=0}$$

$$= \prod^{T} \left(q^{\dagger\prime}_\kappa - \frac{\partial_\ell}{\partial q_\kappa}\right) \prod (q'_\kappa - q_\kappa) \Big|_{q=0} \tag{4.149}$$

$$= \prod \left(1 + q^{\dagger\prime}_\kappa q_\kappa\right) = \langle q^{\dagger\prime}|q'\rangle \quad ,$$

for the opposite sense of multiplication in the two products permits their combination without the intervention of the sign changes that accompany the anticommutativity of the eigenvalues.

The transformation function $\langle q^{\dagger\prime}|q^{\dagger\prime\prime}\rangle$ provides the connection between the wave functions $\psi(q't)$ and $\psi(q^{\dagger\prime}t)$. The composition property

$$\begin{aligned}\langle q^{\dagger\prime}| &= \langle q^{\dagger\prime}|q^{\dagger}\rangle \left\langle \frac{\partial_r^T}{\partial q^{\dagger}} \right| _{q^{\dagger}=0} \\ &= \prod^T (q_\kappa^{\dagger\prime}-q_\kappa^{\dagger}) \left\langle \frac{\partial_r^T}{\partial q^{\dagger}} \right| _{q^{\dagger}=0}\end{aligned} \tag{4.150}$$

yields

$$\psi(q^{\dagger\prime}\ t) = \prod^T (q_\kappa^{\dagger\prime})\ \psi\left(-\frac{\partial_r^T}{\partial q^{\dagger}},\ t\right) \tag{4.151}$$

and similarly

$$\langle q'| = \langle q'|q\rangle \left\langle \frac{\partial_r^T}{\partial q} \right| _{q=0} \tag{4.152}$$

supplies the inverse relation

$$\psi(q't) = \prod (-q'_\kappa)\ \psi\left(-\frac{\partial_r^T}{\partial q'}\ t\right) \quad . \tag{4.153}$$

We also have

$$\phi(q't) = \phi\left(-\frac{\partial_\ell}{\partial q'}\ t\right) \prod (q'_\kappa) \tag{4.154}$$

and

$$\phi(q^{\dagger\prime}t) = \phi\left(-\frac{\partial_\ell}{\partial q^{\dagger\prime}}\ t\right) \prod^T (-q_\kappa^{\dagger\prime}) \quad . \tag{4.155}$$

To obtain an expression of completeness referring entirely to q - eigenvectors, we observe that

$$\begin{aligned} 1 &= |q'\rangle \left\langle \frac{\partial_r^T}{\partial q'} \right| \left. \frac{\partial_\ell}{\partial q''} \right\rangle \langle q''| \ _{q'=q''=0} \\ &= |q'\rangle \prod^T \left(\frac{\partial_r^T}{\partial q'_\kappa} - \frac{\partial_\ell}{\partial q''_\kappa} \right) \langle q''| \ _{q'=q''=0} \quad . \end{aligned} \tag{4.156}$$

If this symbolic form is realized by ϕ wave functions that are commutative with the eigenvalues, or, if n is even, the scalar product evaluation can be presented as

$$\Phi_1\Psi_2 = \prod^T \left(-\frac{\partial_\ell}{\partial q'_\kappa}\right) \phi_1(q't)\ \psi_2(q't)$$
$$= \int d[q']\ \phi_1(q't)\ \psi_2(q't) \quad . \tag{4.157}$$

The integral notation is designed to evoke an analogy and has no significance apart from its differential definition. In a similar way we have

$$1 = |q^{\dagger\prime}> \prod \left(\frac{\partial_\ell}{\partial q^{\dagger\prime\prime}_\kappa} - \frac{\partial^T_r}{\partial q^{\dagger\prime}_\kappa}\right) <q^{\dagger\prime\prime}|\ _{q^{\dagger\prime}=q^{\dagger\prime\prime}=0} \tag{4.158}$$

and

$$\Phi_1\Psi_2 = \prod \left(\frac{\partial_\ell}{\partial q^{\dagger\prime}_\kappa}\right) \phi_1(q^{\dagger\prime}t)\ \psi_2(q^{\dagger\prime}t)$$
$$= \int d[p']\ \phi_1(p't)\ \psi_2(p't) \quad . \tag{4.159}$$

According to the definitions adopted for variables of the second kind,

$$\int d[q']\ \delta[q'-q''] = \prod^T \left(-\frac{\partial_\ell}{\partial q'_\kappa}\right) \prod (q''_\kappa - q'_\kappa) = 1 \tag{4.160}$$

and

$$\int d[p'] \, \delta[p'-p''] = \prod \left(\frac{\partial_\ell}{\partial q_\kappa^{\dagger\prime}} \right) \prod^{T} \left(q_\kappa^{\dagger\prime} - q_\kappa^{\dagger\prime\prime}\right) = 1 \quad . \tag{4.161}$$

Futhermore, the composition property

$$\langle p'|p''\rangle = \int d[q'] \, \langle p'|q'\rangle\langle q'|p''\rangle \tag{4.162}$$

appears as

$$\delta[p'-p''] = \int d[q'] \, e^{-i(p'-p'')q'} \tag{4.163}$$

and similarly

$$\delta[q'-q''] = \int d[p'] \, e^{ip'(q'-q'')} \quad . \tag{4.164}$$

Thus the extension of the delta function notation to the variables of the second kind is not without justification. Another related example of formulae that are applicable to both Hermitian variables of the first kind and the variables of the second kind is obtained from (4.153), written as

$$\psi(q't) = \delta[q'] \, \psi\left(-i \frac{\partial_r^T}{\partial q'} \, t\right) \tag{4.165}$$

with the variables $q^\dagger$ replaced by $p = iq^\dagger$. On inserting the integral representation (4.164) this becomes

$$\psi(q't) = \int d[p'] \, e^{ip'q'} \, \psi(p't) \tag{4.166}$$

and (4.151) supplies the inverse formula

$$\psi(p't) = \int d[q'] \, e^{-ip'q'} \, \psi(q't) \quad . \tag{4.167}$$

For Hermitian variables of the first kind, these are the reciprocal Fourier transformations stated in (4.22).

4.15 INTEGRATION

Although the integral notation is effective in unifying some of the formal properties of the two classes of variables, the nature of the operations is quite distinct. Indeed, the symbol $\int$ has the significance of differentiation for variables of the second kind, and the inverse of differentiation for variables of the first kind. This is emphasized by the effect of subjecting the eigenvalues to a linear transformation , $q' \to \lambda q'$. For the Hermitian variables, the differential element of volume in the q' - space changes in accordance with

$$q_k : \quad d[\lambda q'] = |\det \lambda| \; d[q'] \tag{4.168}$$

which implies that

$$q_k : \quad \delta[\lambda q'] = \frac{1}{|\det \lambda|} \delta[q'] \quad . \tag{4.169}$$

But the delta function of the variables of the second kind is defined as a product of anticommutative factors, and therefore

$$q_K : \quad \delta[\lambda q'] = \det \lambda \; \delta[q'] \tag{4.170}$$

The latter result is also expressed formally by

$$q_K : \quad d[\lambda q'] = \frac{1}{\det \lambda} d[q'] \quad . \tag{4.171}$$

With the particular choice $\lambda=-1$, we learn that $\delta[q']$ is an even function of the Hermitian variables, but it possesses this property for variables of the second kind only if n_2 is even. An interesting formal difference also appears on considering the evaluation, by integration, of the trace of an operator. From the expression of completeness for Hermitian variables of the first kind, (4.23), we derive the integral formulae

$$\text{tr } X = \int d[q'] \; \langle q'|X|q'\rangle = \int d[p'] \; \langle p'|X|p'\rangle \quad . \tag{4.172}$$

To obtain the analogues referring to variables of the second kind, we first note the generally valid differentiation formula, derived from (4.136), in terms of the matrix representation $\langle p'|X|q'\rangle$,

$$\operatorname{tr} X = \left\langle i \frac{\partial_\ell}{\partial q'} \middle| X \middle| q' \right\rangle \Bigg|_{q'=0} = \left\langle p' \middle| X \middle| i \frac{\partial_r^T}{\partial p'} \right\rangle \Bigg|_{p'=0} . \tag{4.173}$$

Then, for variables of the second kind, we deduce $(p = iq^\dagger)$

$$\begin{aligned} \operatorname{tr} X &= \left\langle \frac{\partial_\ell}{\partial q'} \middle| \frac{\partial_\ell}{\partial q''} \right\rangle \langle q''|X|q'\rangle \Bigg|_{q'=q''=0} \\ &= \prod^T \left(\frac{\partial_\ell}{\partial q'} - \frac{\partial_\ell}{\partial q''} \right) \langle q''|X|q'\rangle \Bigg|_{q'=q''=0} \\ &= \int d[q'] \, \langle q'|X|-q'\rangle \quad , \end{aligned} \tag{4.174}$$

and

$$\begin{aligned} \operatorname{tr} X &= \langle q^{\dagger\prime}|X|q^{\dagger\prime\prime}\rangle \left\langle \frac{\partial_r^T}{\partial q^{\dagger\prime\prime}} \middle| \frac{\partial_r^T}{\partial q^{\dagger\prime}} \right\rangle \Bigg|_{q^{\dagger\prime}=q^{\dagger\prime\prime}=0} \\ &= \langle q^{\dagger\prime}|X|q^{\dagger\prime\prime}\rangle \prod \left(\frac{\partial_r^T}{\partial q^{\dagger\prime}} - \frac{\partial_r^T}{\partial q^{\dagger\prime\prime}} \right) \Bigg|_{q^{\dagger\prime}=q^{\dagger\prime\prime}=0} \\ &= \int d[p'] \, \langle p'|X|-p'\rangle \quad . \end{aligned} \tag{4.175}$$

With even n_2 we have, for example

$$\rho|q'\rangle = \rho e^{-ipq'}|0\rangle = e^{ipq'}\rho|0\rangle$$
$$= |-q'\rangle \tag{4.176}$$

and these trace formulae can also be expressed as

$$\text{tr } X = \int d[q'] \langle q'|X\rho|q'\rangle = \int d[p'] \langle p'|X\rho|p'\rangle \quad , \tag{4.177}$$

which restores somewhat the uniformity of the two classes for it is the unit operator, commuting with all variables, that plays the role of ρ for variables of the first kind. Since the trace is unaltered on replacing X with $\rho X \rho$, one can aso write ρX in place of $X\rho$, which is to say that odd functions of the variables have vanishing trace. The simplest example of a trace evaluation is

$$\text{tr } 1 = \int d[q'] \langle q'|-q'\rangle$$
$$= \prod^{T} \left(- \frac{\partial_\ell}{\partial q'}\right) \prod \left(-2q'\right) = 2^{n_2} \quad . \tag{4.178}$$

Of course, with a system requiring both types of dynamical variables for its description, the trace operations referring to the two classes must be superimposed. A formula of suitable generality for even n_2 , derived from (4.172), (4.174), and (4.166), is

$$\text{tr}\ X = \int d[q']\, d[p']\, e^{iq'p'} \langle p'|X|q'\rangle \quad . \tag{4.179}$$

4.16 DIFFERENTIAL REALIZATIONS

Finally, we note the universality of the differential operator realizations

$$\begin{aligned} \langle p't|q(t) &= i \frac{\partial_\ell}{\partial p'} \langle p't| \quad , \\ p(t)|q't\rangle &= |q't\rangle\, i \frac{\partial_r^T}{\partial q'} \quad , \end{aligned} \tag{4.180}$$

with their algebraic generalizations

$$\begin{aligned} \langle p't|F(qp) &= F\left(i \frac{\partial_\ell}{\partial p'} , p'\right) \langle p't| \\ F(qp)|q't\rangle &= |q't\rangle\, F\left(q' , i \frac{\partial_r^T}{\partial q'}\right) \quad , \end{aligned} \tag{4.181}$$

which imply the Schrödinger differential equations

$$i \frac{\partial}{\partial t} \psi(p't) = H\left(i \frac{\partial_\ell}{\partial p'} , p' , t\right) \psi(p't) \tag{4.182}$$

and

$$\phi(q't)\,\frac{1}{i}\,\frac{\partial^T}{\partial t} = \phi(q't)\,H\!\left(q'\,,\,i\,\frac{\partial_r^T}{\partial q'}\,,\,t\right) \quad . \tag{4.183}$$

The $\langle q't|$ and $|p't\rangle$ states do not permit such general assertions, for, in addition to the asymmetry characteristic of the non-Hermitian variables of the first kind, we must distinguish between the two classes of variables in the differential operator realizations

$$\begin{aligned} \langle q't|F(qp) &= F\!\left(q'\,,\,\mp\, i\,\frac{\partial_\ell}{\partial q'}\right)\langle q't| \\ F(qp)|p't\rangle &= |p't\rangle\,F\!\left(\mp\, i\,\frac{\partial_r^T}{\partial p'}\,,\,p'\right) \quad , \end{aligned} \tag{4.184}$$

where the upper sign refers to variables of the first kind. This sign distinction originates from the necessary change of multiplication order in the differential expression

$$\begin{aligned} \delta_q\langle q't| &= i\langle q't|p\,\delta q' = i\langle q't|(\pm)\,\delta q'\,p \\ &= (\pm)\,i\,\delta q'\,\langle q't|p \quad , \end{aligned} \tag{4.185}$$

as contrasted with

$$\delta_p\langle p't| = -\,i\langle p't|\,\delta p'\,q = -i\,\delta p'\,\langle p't|q \quad . \tag{4.186}$$

Thus the following Schrödinger equations are applicable to Hermitian variables of the first kind and to the variables of the second kind,

$$i \frac{\partial}{\partial t} \psi(q't) = H\left(q', \mp i \frac{\partial_\ell}{\partial q'}, t\right) \psi(q't)$$

$$\phi(p't) \frac{1}{i} \frac{\partial^T}{\partial t} = \phi(p't) \, H\left(\mp i \frac{\partial_r^T}{\partial p'}, p', t\right) \quad . \tag{4.187}$$

CHAPTER FIVE
CANONICAL TRANSFORMATIONS

The use of a canonical version of the Lagrangian operator, such as

$$L = p \cdot \frac{dq}{dt} - H \quad , \tag{5.1}$$

yields the corresponding canonical form of the generator describing infinitesimal transformations at a specified time t ,

$$G = p\,\delta q - H\,\delta t \quad , \tag{5.2}$$

which implies the canonical commutation relations and equations of motion. From a given generator G other generators $\bar{G}$ can be obtained, in accordance with

$$G - \bar{G} = \delta W \quad , \tag{5.3}$$

and we now ask whether $\bar{G}$ can also be exhibited in a canonical form referring to new dynamical variables $\bar{q}(t)$, $\bar{p}(t)$, and a new Hamiltonian operator $\bar{H}(\bar{q}\bar{p}t)$. Such new variables would then obey canonical commutation relations and equations of motion, characterizing the transformation of dynamical variables at the time t as a canonical transformation. The differential form

$$\delta W(q\bar{q}t) = p\,\delta q - H\,\delta t - \bar{p}\delta\bar{q} + \bar{H}\,\delta t \tag{5.4}$$

appears to be the result of subjecting the variables in the action operator W to infinitesimal variations, which for q and $\bar{q}$, are the special operators variations. However, the action operator can also contain other, superfluous, canonical variables v , and we infer the differential equations

$$p = \frac{\partial_r W}{\partial q} \quad , \quad -\bar{p} = \frac{\partial_r W}{\partial \bar{q}} \quad , \quad \frac{\partial W}{\partial v} = 0 \tag{5.5}$$

$$\bar{H} = H + \frac{\partial W}{\partial t} \quad .$$

A canonical transformation is obtained should these implicit operator equations possess a solution for $\bar{q}$ and $\bar{p}$.

5.1 GROUP PROPERTIES AND SUPERFLUOUS VARIABLES

Canonical transformations form a group. The action operator that describes the transformation inverse to $qp \to \bar{q}\bar{p}$ is

$$W(\bar{q} \, , \, q) = -W(q \, , \, \bar{q}) \tag{5.6}$$

while, for two successive transformations, $qp \to \bar{q}\bar{p} \to \bar{\bar{q}}\bar{\bar{p}}$, the generating action operator of the composite transformation is

$$W(q \, , \, \bar{\bar{q}}) = W(q \, , \, \bar{q}) + W(\bar{q} \, , \, \bar{\bar{q}}) \quad . \tag{5.7}$$

The latter form illustrates the concept of superfluous variable. If the individual operators $W(q \, , \, \bar{q})$ and $W(\bar{q} \, , \, \bar{\bar{q}})$ contain just the indicated variables, the sum (5.7) involves q, $\bar{\bar{q}}$ and $\bar{q}$.

But, according to the statements of the individual canonical transformations,

$$\frac{\partial r}{\partial \bar{q}} [W(q , \bar{q}) + W(\bar{q} , \bar{\bar{q}})] = - \bar{p} + \bar{p} = 0 \qquad (5.8)$$

and it must be possible to exhibit $W(q , \bar{\bar{q}})$ as a function of the variables $q , \bar{\bar{q}}$ only. It is not always desirable, however, to eliminate the superfluous variables. This is particularly true when the canonical transformation involves algebraic relations between the variables q and $\bar{\bar{q}}$, which inhibit their independent variation. By retaining the variables of a suitable intermediate transformation, one can derive the desired transformation by independent differentiation. An important example is provided by the identity transformation. Let us remark first on the now familiar transformation that interchanges the roles of the complementary q and p variables, as described by

$$G_q - G_p = \delta W(q , p) \qquad (5.9)$$

where

$$W(q , p) = - W(p , q) = p.q \quad . \qquad (5.10)$$

Since

$$G_p = -\delta p\, q = -\sum_k q_k\, \delta p_k + \sum_\kappa q_\kappa\, \delta p_\kappa \qquad (5.11)$$

this is the canonical transformation

$$\begin{aligned} \bar{q}_k &= p_k \ , \quad \bar{p}_k = -q_k \\ \bar{q}_\kappa &= \bar{p}_\kappa \ , \quad \bar{p}_\kappa = q_\kappa \ . \end{aligned} \qquad (5.12)$$

On adding the action operators for the transformation $q \rightarrow p$ and its inverse $p \rightarrow q = \bar{q}$, we obtain the following characterization of the identity transformation,

$$W(q\ ,\ \bar{q}) = p_{\bullet}(q-\bar{q}) \ , \qquad (5.13)$$

which contains the p as superfluous variables. To eliminate the latter we must impose the transformation equations $\bar{q}=q$, which yields $W=0$ for the identity transformation. But, with the superfluous variables retained, the differential equations (5.5) are applicable and generate the transformation.

A given canonical transformation can be derived by differentiation with respect to either set of complementary variables. Thus, from the action operator $W(q\bar{q}t)$, obeying the differential equations

(5.5), we obtain

$$\begin{aligned} W(p\bar{q}t) &= W(p\ ,\ q) + W(q\bar{q}t) \\ &= -\ p_{\bullet}q + W(q\bar{q}t)\quad , \end{aligned} \tag{5.14}$$

in which the q appear as superfluous variables. The differential properties of the new action operator $W(p\bar{q}t)$ are now deduced to be

$$-q = \frac{\partial_{\ell} W}{\partial p}\quad , \quad -\bar{p} = \frac{\partial_{r} W}{\partial \bar{q}}\quad , \quad \frac{\partial W}{\partial v} = 0$$
$$\bar{H} = H + \frac{\partial W}{\partial t}\quad , \tag{5.15}$$

and these equations are equally suitable for describing the transformation $qp \to \bar{q}\bar{p}$. The identity transformation, for example, is derived from the action operator $-\ p\ \bar{q}$.

5.2 INFINITESIMAL CANONICAL TRANSFORMATIONS

Transformations in the infinitesimal neighborhood of the identity - infinitesimal canonical transformations - must be described by an action operator that differs infinitesimally from the one producing the identity transformation. The appropriate form for the variables q , $\bar{q}$ is

$$W(q\bar{q}t) = p_{\bullet}(q-\bar{q}) - G(\bar{q}pt) \tag{5.16}$$

where G is an infinitesimal function of the indicated variables that should have an even dependence upon the second class of dynamical variable, but is otherwise arbitrary. On applying (5.5), with the p as superfluous variables, we obtain the explicit equations of an infinitesimal canonical transformation,

$$\bar{q} = q - \delta q \quad , \quad \bar{p} = p - \delta p \quad , \tag{5.17}$$

with

$$\begin{aligned} \delta q &= \frac{\partial_\ell}{\partial p} G(qpt) \quad , \quad \delta p = -\frac{\partial_r}{\partial q} G(qpt) \\ \bar{H}(\bar{q}\bar{p}t) &= H(qpt) - \frac{\partial}{\partial t} G(qpt) \quad . \end{aligned} \tag{5.18}$$

Now, according to (3.100) and (3.123), which applies to even functions of the variables of the second kind, we have

$$\delta q = \frac{1}{i}[q , G] \quad , \quad \delta p = \frac{1}{i}[p , G] \tag{5.19}$$

or, without specialization to canonical variables,

$$\delta x = \frac{1}{i}[x , G] \quad . \tag{5.20}$$

Hence

$$\bar{x} = x - \delta x = (1 - iG)x(1 + iG) \quad , \qquad (5.21)$$

and if G is a Hermitian operator this is a unitary transformation. The subgroup of canonical transformations that preserve the Hermiticity of dynamical variables is equivalent to the group of unitary transformations. Without reference to Hermitian properties the transformation (5.21) maintains all algebraic relations, and therefore

$$\begin{aligned} \bar{H}(\bar{x}t) &= (1 - iG)\bar{H}(xt)(1 + iG) \\ &= \bar{H}(xt) + \frac{1}{i}[G , H] \quad , \end{aligned} \qquad (5.22)$$

whence

$$\begin{aligned} \bar{H}(xt) - H(xt) &= -\frac{\partial}{\partial t} G - \frac{1}{i}[G , H] \\ &= -\frac{d}{dt} G \quad , \end{aligned} \qquad (5.23)$$

which makes explicit the functional form of the new Hamiltonian. Infinitesimal canonical transformations that do not change the form of the Hamiltonian operator have infinitesimal generators that are constants of the motion.

We have already encountered examples of infinitesimal canonical transformations. The tranformation generated by $G_t = - H\,\delta t$,

$$\delta x = \frac{1}{i}\,[x\;,\;-H\,\delta t] = -\,\frac{dx}{dt}\,\delta t \qquad (5.24)$$

is one in which the dynamical variables at time t are replaced by those at time $t + \delta t$,

$$\bar{x}(t) = x - \delta x = x(t + \delta t) \quad . \qquad (5.25)$$

The new form of the Hamiltonian operator is

$$\bar{H}(\bar{x}t) = H\big(x(t + \delta t)\;,\;t + \delta t\big) = H(\bar{x}t) + \frac{\partial H}{\partial t}\,\delta t \quad , \qquad (5.26)$$

and the energy operator H is a constant of the motion when H is not an explicit function of t , being the condition for the maintenance of the functional form of the Hamiltonian operator under time translation. The special operator variations, which are distinguished by their elementary commutation properties, appear as the canonical transformations

$$\delta q = \frac{\partial_\ell G_q}{\partial p} = \delta_s q \quad , \qquad \delta p = -\,\frac{\partial_r G_q}{\partial q} = 0 \qquad (5.27)$$

and

$$\delta q = \frac{\partial_\ell G_p}{\partial p} = 0 \quad , \quad \delta p = - \frac{\partial_r G_p}{\partial p} = \delta_s p \quad , \qquad (5.28)$$

in which we have introduced a notational distinction between the general infinitesimal canonical transformations and those of the special canonical group. For variables of the second kind, the consideration of the special canonical group within the framework of general canonical transformations implies a formal extension of the latter through the introduction of the elements of the external algebra.

5.3 ROTATIONS. ANGULAR MOMENTUM

The change in description that accompies a rotation of the spatial coordinate system is a canonical transformation, with the infinitesimal generator

$$G_{\boldsymbol{\omega}} = \delta\boldsymbol{\omega} \cdot \mathbf{J} \quad , \qquad (5.29)$$

but the form of this canonical transformation is not yet known. Differently oriented coordinate systems are intrinsically equivalent and we should expect that the kinematical term in the Lagrangian operator presents the same appearance in terms of the variables appropriate to any coordinate system. This comment also applies to the dynamical term - the Hamiltonian operator - of a physically isolated system, for which the total angular momentum operator $\mathbf{J}$, as the generator of a transformation that leaves the form of H invariant, is a constant of the motion. In view of the bilinear structure of the kinematical term, $p \cdot \frac{dq}{dt}$, the change involved in an arbitrary rotation of the spatial coordinate system will be a linear transformation among suitably chosen q variables, combined with the contragredient transformation of the complementary variables. For an infinitesimal rotation, then,

$$\delta_{\boldsymbol{\omega}} q = \frac{\partial_\ell G_{\boldsymbol{\omega}}}{\partial p} = -i\, \delta\boldsymbol{\omega} \cdot \mathbf{j}\, q$$
$$\delta_{\boldsymbol{\omega}} p = -\frac{\partial_r G_{\boldsymbol{\omega}}}{\partial q} = i\, p\, \delta\boldsymbol{\omega} \cdot \mathbf{j} \quad , \tag{5.30}$$

in which the components of the vector $\mathbf{j}$ are matrices. For Hermitain variables $\mathbf{j}$ is an imaginary matrix, while with non Hermitian variables in the relation $p = iq^\dagger$, the matrix $\mathbf{j}$ is

Hermitian. The general form of the Hermitian angular momentum operator is thus obtained as

$$\mathbf{J} = -\, ip_{\bullet}jq \quad , \tag{5.31}$$

to which the two kinds of dynamical variables make additive contributions. The symmetrization or antisymmetrization indicated here **is actually** unnecessary.

On applying the infinitesimal transformation generated by $G_{\boldsymbol{\omega}}$ to the operator $\mathbf{J}$ we find

$$\delta_{\boldsymbol{\omega}}\mathbf{J} = \frac{1}{i}\,[\mathbf{J}\ ,\ \ \delta\boldsymbol{\omega}\cdot\mathbf{J}\,] = \delta\boldsymbol{\omega}\times\mathbf{J} \quad , \tag{5.32}$$

according to the commutation relations (2.79). But we also have

$$\begin{aligned}\delta_{\boldsymbol{\omega}}\mathbf{J} &= -\, i[pj\ \delta_{\boldsymbol{\omega}}\, q + \delta_{\boldsymbol{\omega}}p\ jq] \\ &= -\, ip\,\frac{1}{i}\,[\mathbf{j}\ ,\ \delta\boldsymbol{\omega}\cdot\mathbf{j}\,]q\end{aligned} \tag{5.33}$$

and therefore

$$\frac{1}{i}\,[\mathbf{j}\ ,\ \ \delta\boldsymbol{\omega}\cdot\mathbf{j}\,] = \delta\boldsymbol{\omega}\times\mathbf{j} \quad ; \tag{5.34}$$

the matrices $\mathbf{j}$ obey the same commutation relations as the angular momentum operator $\mathbf{J}$. It is a consequence of these commutation relations that the trace of the matrix $\mathbf{j}$ vanishes and thus the explicit symmetrization or antisymmetrization of factors in $\mathbf{J}$ is unnecessary. The decompositon of the matrix j into irreducible submatrices produces a partitioning of the dynamical variables into kinematically independent sets that appear additively in the structure of $\mathbf{J}$. Each such set defines a dynamical variable of several components, the rotational transformation properties of which are fixed by the number of components, for this integer, the dimensionality of the corresponding submatrix of j, essentially determines the structure of these matrix representations of $\mathbf{J}$. An irreducible set of three variables, for example, necessarily has the rotational transformation properties of a three-dimensional space vector. The number of components possessed by a dynamical variable of the second kind is presumably even, according to a comment of the preceding section.

5.4 TRANSLATIONS. LINEAR MOMENTUM

The remarks concerning invariance with respect to rotations of the coordinate system apply equally to coordinate system translations, which have the infinitesimal generator

$$G_{\varepsilon} = \delta\boldsymbol{\varepsilon} \cdot \mathbf{P} \tag{5.35}$$

For systems described by a finite number of dynamical variables, the appropriate transformation that leaves the kinematical part of L invariant is the addition of constants to suitably chosen canonical variables q. Thus G_ε has the structure of G_q and different generators of this type do commute, as the commutation properties of the total linear momentum require. If we apply an infinitesimal rotation to the operator

$$\mathbf{P} = p \cdot \frac{\delta q}{\delta \boldsymbol{\varepsilon}} , \tag{5.36}$$

in accordance with

$$\begin{aligned} \delta_{\boldsymbol{\omega}} \mathbf{P} &= \frac{1}{i} [\mathbf{P} , \ \delta\boldsymbol{\omega} \cdot \mathbf{J}] = \delta\boldsymbol{\omega} \times \mathbf{P} \\ &= \delta_{\boldsymbol{\omega}} p \cdot \frac{\delta q}{\delta \boldsymbol{\varepsilon}} , \end{aligned} \tag{5.37}$$

we recognize that the members of the class of dynamical variables p that make a contribution to $\mathbf{P}$ have the rotational transformation properties of space vectors. Hence, only three-component variables of the first kind can be affected by a translation of the coordinate system. If the latter set of variables is presented as the Hermitian vectors

$\mathbf{r}_1, \ldots, \mathbf{r}_n$, with the complementary variables $\mathbf{p}_1, \ldots, \mathbf{p}_n$, a suitable adjustment of the relative eigenvalue scales will guarantee that

$$\delta_{\boldsymbol{\varepsilon}} \mathbf{r}_k = \delta \boldsymbol{\varepsilon} \tag{5.38}$$

and therefore

$$\mathbf{P} = \sum_{k=1}^{n} \mathbf{p}_k \quad . \tag{5.39}$$

One exhibiting the contribution of the vector variables to the total angular momentum, we have

$$\mathbf{J} = \sum_{k=1}^{n} \mathbf{r}_k \times \mathbf{p}_k + \mathbf{J}_{int} \tag{5.40}$$

where the latter term contains all variables that are uninfluenced by translations. These are evidently the internal variables for a system of n particles that are localized spatially by the position vectors $\mathbf{r}_k$.

5.5 TRANSFORMATION PARAMETERS

It is useful to regard a general infinitesimal canonical transformation as the result of subjecting certain parameters τ_s , $s = 1,\ldots,\nu$, to infinitesimal changes, say $-d\tau_s$, so that the infinitesimal generator has the form

$$G = \sum_{s=1}^{\nu} G_{(s)}(-d\tau_s) \quad , \tag{5.41}$$

in which the $G_{(s)}$ may depend explicitly upon the parameters. This interpretation of the transformation is expressed by

$$\bar{x} = x + \sum \frac{dx}{d\tau_s}(-d\tau_s) \tag{5.42}$$

or

$$\delta x = x - \bar{x} = \sum \frac{dx}{d\tau_s} d\tau_s = dx \quad . \tag{5.43}$$

Accordingly, the canonical variables obey equation of motion,

$$\frac{dq}{d\tau_s} = -\frac{\partial_\ell G_{(s)}}{\partial p} \quad , \quad \frac{dp}{d\tau_s} = \frac{\partial_r G_{(s)}}{\partial q} \quad , \tag{5.44}$$

which govern the evolution of the canonical transformation. By repeated application of such infinitesimal transformations, a finite transformation is generated in which the parameters τ are altered from τ_1 to τ_2 along a definite path. The action operator characterizing the finite transformation is the sum of those for the individual infinitesimal transformations

$$W_{12} = \sum_{\tau_2}^{\tau_1} W_{\tau,\ \tau-d\tau} \tag{5.45}$$

where, according to (5.16),

$$W_{\tau,\ \tau-d\tau} = p(\tau) . \big(q(\tau) - q(\tau-d\tau)\big) + \sum_s G_{(s)} \big(q(\tau-d\tau)\ ,\ p(\tau)\ ,\ \tau\big)\ d\tau_s\ . \tag{5.46}$$

Hence

$$W_{12} = \int_{\tau_2}^{\tau_1} \big(p.dq + \ \Sigma\ G_{(s)}\ d\tau_s\big) \tag{5.47}$$

is the action operator generating the finite canonical transformation, with the operators referring to all values of τ intermediate between τ_1 and τ_2 appearing as superfluous variables. Since W_{12} must be independent of these intermediate variables, it is stationary with respect to infinitesimal special variations of all dynamical quantities that do not refer to the terminal values of

τ . Now

$$\delta\,[p\,.\,dq + \Sigma\, G_{(s)}\, d\tau_s] - d\,[p\,\delta q + \Sigma\, G_{(s)}\, \delta\tau_s]$$
$$= \delta p\, dq - dp\, \delta q + \sum \left(\delta G_{(s)}\, d\tau_s - dG_{(s)}\, \delta\tau_s\right) \tag{5.48}$$

which in turn, equals

$$\delta p \left(dq + \sum \frac{\delta_\ell G_{(s)}}{\delta p} d\tau_s \right) - \left(dp - \sum \frac{\delta_r G_{(s)}}{\delta q} d\tau_s \right) \delta q$$
$$+ \sum_{rs} \left(\frac{\partial G_{(s)}}{\partial \tau_r} - \frac{dG_{(r)}}{d\tau_s} \right) \delta\tau_r\, d\tau_s \quad , \tag{5.49}$$

and thus the stationary requirement, applied to a given parameter path $(\delta\tau = 0)$, again yields the differential equations (5.44), and

$$\delta W_{12} = G_1 - G_2$$
$$G = p\,\delta q + \sum G_{(s)}\, \delta\tau_s \quad . \tag{5.50}$$

5.6 HAMILTON - JACOBI TRANSFORMATION

With a single parameter t and generator $-H$, we regain the original action principle, now appearing as the characterization of a canonical transformation - the Hamiltonian - Jacobi transformation - from a description at time t $(=t_1)$ to the analogous one referring to another time t_0 $(=t_2)$. With t_0 held fixed, the action operator W $(=W_{12})$ obeys

$$\delta W = p\ \delta q - H\ \delta t - p_0\ \delta q_0 \tag{5.51}$$

or

$$p = \frac{\partial_r W}{\partial\ q}\ , \quad -p_0 = \frac{\partial_r W}{\partial q_0} \qquad H + \frac{\partial W}{\partial t} = 0\ . \tag{5.52}$$

On comparison with (5.5) we recognize that the Hamilton-Jacobi transformation is such that $\bar{H} = 0$, which expresses the lack of dependence on t of the new dynamical variables $x(t_0)$. The new Hamiltonian at time t thus differs from H evaluated at time t_0 , which governs the dependence of W upon the parameter t_0 ,

$$\frac{\partial W}{\partial t_0} = H(x(t_0)\ , t_0)\ . \tag{5.53}$$

5.7 PATH DEPENDENCE

For a canonical transformation involving several parameters, the last term of (5.49) displays the effect of altering the path along which the parameters evolve. According to the significance of $G_{(s)}$ as a generator, we have

$$\frac{dG_{(r)}}{d\tau_s} = \frac{\partial G_{(r)}}{\partial \tau_s} - \frac{1}{i}\,[G_{(r)}\ ,\ G_{(s)}] \tag{5.54}$$

and therefore

$$\frac{\partial G_{(s)}}{\partial \tau_r} - \frac{dG_{(r)}}{d\tau_s} = R_{rs} \tag{5.55}$$

is antisymmetrical with respect to the indices r and s ,

$$\begin{aligned} R_{rs} &= \frac{\partial G_{(s)}}{\partial \tau_r} - \frac{\partial G_{(r)}}{\partial \tau_s} + \frac{1}{i}\,[G_{(r)}\ ,\ G_{(s)}] \\ &= -\,R_{sr}\ . \end{aligned} \tag{5.56}$$

The complete variation of W_{12} is thus

$$\delta W_{12} = G_1 - G_2 + \int_{\tau_2}^{\tau_1} \sum_{rs} R_{rs}\, \tfrac{1}{2}\left(\delta\tau_r\, d\tau_s - \delta\tau_s\, d\tau_r\right) . \tag{5.57}$$

With fixed terminal conditions, the consideration of two independent path variations in the combinat-ion

$$\left(\delta^{(1)} \delta^{(2)} - \delta^{(2)} \delta^{(1)} \right) W_{12}$$

$$\int_{\tau_1}^{\tau_2} \sum \left[\frac{dR_{rs}}{d\tau_q} + \frac{dR_{sq}}{d\tau_r} + \frac{dR_{qr}}{d\tau_s} \right] \tag{5.58}$$

$$\times \frac{1}{2} \left(\delta^{(1)}\tau_q \, \delta^{(2)}\tau_r - \delta^{(2)}\tau_q \, \delta^{(1)}\tau_r \right) d\tau_s \quad ,$$

leads to the integrability condition of the differential form (5.57) in its dependence upon the parameter path,

$$\frac{dR_{rs}}{d\tau_q} + \frac{dR_{sq}}{d\tau_r} + \frac{dR_{qr}}{d\tau_s} = 0 \quad , \tag{5.59}$$

which is indeed satisfied by virtue of the operator identity (2.73), for

$$\frac{dR}{d\tau_q} = \left[\frac{\partial}{\partial\tau_q} - iG_{(q)} \, , \, R \right] \tag{5.60}$$

and

$$R_{rs} = i \left[\frac{\partial}{\partial\tau_r} - iG_{(r)} \, , \, \frac{\partial}{\partial\tau_s} - iG_{(s)} \right] . \tag{5.61}$$

5.8 PATH INDEPENDENCE

If the canonical transformation is to be independent of the integration path, it is necessary that

$$R_{rs} = 0 \quad . \tag{5.62}$$

As an example of this situation, consider a canonical transformation with two sets of parameters and generators: t , $-H$; λ , $G_{(\lambda)}$. On referring to (5.55) we see that the condition for path independence can be presented as

$$\frac{\partial H}{\partial \lambda} = - \frac{dG_{(\lambda)}}{dt} \quad . \tag{5.63}$$

Hence the Hamiltonian operator must be an explicit function of the parameter λ , which is to say that it changes its functional form under the infinitesimal transformation $G_{(\lambda)}\, d\lambda$. Since this change is identical with (5.23), characteristic of an infinitesimal canonical transformation, we learn that the same resultant canonical transformation is obtained whether the system evolves in time and a canonical transformation is performed at the terminal time, or if the canonical transformation is applied continuously in time, subject only to the fixed endpoint. With the latter viewpoint, the superposition of the continuous change in description on the dynamical development of the system is described by the effective Hamiltonian operator

$$H_e = H(\lambda) - G_{(\lambda)} \frac{d\lambda}{dt} \quad , \tag{5.64}$$

and the principle of stationary action includes the numerical variable $\lambda(t)$. It should be noted that when $G_{(\lambda)}$ is not an explicit function of time, the λ dependence of H is such that $dH/d\lambda = 0$.

5.9 LINEAR TRANSFORMATIONS

Under some circumstances, this extension of the action principle can be expressed as a widening of the class of variations, without alteration of the Hamiltonian. Thus, let

$$G_{(\lambda)} = - ip.gq \tag{5.65}$$

which produces the linear transformation described by

$$\frac{dq}{d\lambda} = igq \quad , \quad -\frac{dp}{d\lambda} = ipg \quad . \tag{5.66}$$

If g is a constant matrix, the explicit λ transformation, for constant t, is

$$g(\lambda) = e^{i\lambda g}q \quad , \quad p(\lambda) = pe^{-i\lambda g} \quad . \tag{5.67}$$

If these linear operator relations are substituted into the Lagrangian operator, we obtain

$$p(\lambda) \cdot \frac{dq(\lambda)}{dt} + G_{(\lambda)} \frac{d\lambda}{dt} - H\big(q(\lambda) \; , \; p(\lambda) \; , \; \lambda\big)$$

$$= p \cdot \frac{dq}{dt} - H \quad , \tag{5.68}$$

from which all reference to the λ transformation has disappeared. We have used the properties

$$p(\lambda) \cdot \frac{dq(\lambda)}{dt} = p \cdot \frac{dq}{dt} \quad , \tag{5.69}$$

for fixed λ, and

$$G_{(\lambda)} = p \cdot \frac{\partial G_{(\lambda)}}{\partial p} = -\, p(\lambda) \cdot \frac{dq(\lambda)}{d\lambda} \quad . \tag{5.70}$$

The action operator is thereby expressed as that of a pure Hamilton-Jacobi transformation. But the λ transformation can now be introduced by remarking that a special variation of $q(\lambda)$ and an infinitesimal change of λ, in

$$q = e^{-i\lambda g} q(\lambda) \tag{5.71}$$

implies a special variation of q, combined with a linear transformation:

$$\delta q - i \, \delta\lambda \, gq \quad . \tag{5.72}$$

Together with the similar properties of p, this yields an extended class of variations for the action principle. To verify directly the correctness of this extension, we observe that the latter induces the following variation in L,

$$\begin{aligned} & G_{(\lambda)} \frac{d \, \delta\lambda}{dt} - \frac{\partial H}{\partial \lambda} \, \delta\lambda \\ &= \frac{d}{dt} \left[G_{(\lambda)} \, \delta\lambda \right] - \left(\frac{\partial H}{\partial \lambda} + \frac{dG_{(\lambda)}}{dt} \right) \delta\lambda \quad , \end{aligned} \tag{5.73}$$

in which $\partial H/\partial\lambda$ has been introduced to measure the lack of invariance displayed by H under the λ transformation. The application of the stationary action principle now properly yields (5.63) and confirms the interpretation of $G_{(\lambda)} \, \delta\lambda$ as the generator of the infinitesimal λ transformation.

We may well note here the special situation of the linear Hamilton-Jacobi transformation, corresponding to the bilinear Hamiltonian operator

$$H = -\, ip.hq \tag{5.74}$$

and the equations of motion

$$\frac{dq}{dt} = -\, ihq \quad , \quad \frac{dp}{dt} = pih \quad . \tag{5.75}$$

Since

$$p \cdot \frac{dq}{dt} = H \quad , \tag{5.76}$$

in virtue of the equations of motion, the action operator $W(qq_0t)$ is identically zero, which indicates the existence of algebraic relations between the variables q and q_0 . The transformation is more conveniently described with the aid of $W(pq_0t)$. According to (5.14), we must eliminate the variables q , which is accomplished by the explicit solution of the equations of motion,

$$q = e^{-ih(t-t_0)}\, q_0 \tag{5.77}$$

(h constant) , and thus

$$W(p\ ,\ q_0\ ,\ t-t_0) = -\ p.q \tag{5.78}$$

$$= -\ p.e^{-ih(t-t_0)}\ q_0\ .$$

CHAPTER SIX

GROUPS OF TRANSFORMATIONS

6.1 INTEGRABILITY CONDITIONS

We will now examine the construction from its infinitesimal elements of a continuous group of canonical transformations, where a transformation must be completely specified by the values of the parameters and thus is independent of the integration path. Apart from the elementary situation of a completely commutative (Abelian) group, the generators $G_{(s)}$ must be explicit funtions of the parameters if the operators R_{rs} are to be zero. The group property is exploited in exhibiting the operators $G_{(s)}(x, \tau)$, assumed finite in number, as a linear combination of an equal number of operators that do not depend explicity upon the parameters,

$$G_{(s)}(x, \tau) = \sum_a G_{(a)}(x)\, \xi_{as}(\tau) \quad , \qquad (6.1)$$

for the condition of path independence demands that the commutators of the operators $G_{(a)}(x)$ be linearly related to the same set. On writing

$$[G_{(b)}, G_{(c)}] = \sum_a G_{(a)}\, g_{abc} \quad , \qquad (6.2)$$

we obtain the following differential equations for the functions $\xi_{as}(\tau)$

$$\frac{\partial}{\partial\tau_r}\xi_{as} - \frac{\partial}{\partial\tau_s}\xi_{ar} = i\sum_{bc} g_{abc}\,\xi_{br}\,\xi_{cs} \quad . \tag{6.3}$$

The numbers g_{abc} are antisymmetrical in the last two indices, and they are imaginary if the operators $G_{(a)}$ are Hermitian. Other algebraic properties can be conveniently presented by introducing a matrix notation for the array with fixed second index,

$$g_b = (g_{abc}) \quad , \tag{6.4}$$

while writing the commutation relations (6.2) in the form

$$[G_{(b)}\ , G] = G\,g_b \quad . \tag{6.5}$$

The latter establishes a correspondence between the operator $G_{(b)}$ and the finite matrix g_b . This correspondence maintains commutation properties according to the identity (2.73),

$$\begin{aligned}&[G_{(a)}\ , [G_{(b)}\ , G]] - [G_{(b)}\ , [G_{(a)}\ , G]]\\&= [[G_{(a)}\ , G_{(b)}]\ , G] = G[g_a\ , g_b] \quad ,\end{aligned} \tag{6.6}$$

and thus the g matrices also obey the commutation relations (6.2). A second set of matrices with that property, $-g^T$, follows from the correspondence

$$[G\ ,\ G_{(b)}] = -g_b^T\ G\ . \tag{6.7}$$

The quadratic connections among the g coefficients, comprised in the commutation relations, are identical with the conditions of integrability for the differential equations (6.3), which verifies the consistency of the operator presentation (6.1).

6.2 FINITE MATRIX REPRESENTATION

The correspondence between the operators $G_{(a)}$ and the matrices g_a persists under a change of operator basis, to within the freedom of matrix transformations that preserve algebraic relations; the non-singular transformation

$$\bar{G}_{a'} = \sum_a G_a\ \lambda_{aa'} \tag{6.8}$$

induces

$$\bar{g}_{a'} = \lambda^{-1} \left(\sum_{a} g_a \lambda_{aa'} \right) \lambda \quad . \tag{6.9}$$

An important application of this ability to change the operator basis occurs when the operators G are Hermitian and possess a non-zero, linearly independent, finite-dimensional, Hermitian matrix representation. Then, with the trace computed from the bounded matrix representation,

$$\mathrm{tr}\,\left([G_{(a)}\,,\,G_{(b)}]G_{(c)}\right) = \mathrm{tr}\,\left([G_{(b)}\,,\,G_{(c)}]G_{(a)}\right)$$
$$= \mathrm{tr}\,\left([G_{(c)}\,,\,G_{(a)}]G_{(b)}\right) = \sum_{a'} \gamma_{aa'}\, g_{a'bc} \tag{6.10}$$

is completely antisymmetrical in a , b , and c , where

$$\gamma_{aa'} = \mathrm{tr}\, G_{(a)}G_{(a')} \tag{6.11}$$

is a real, symmetric, positive-definite matrix. Accordingly, there is a choice of Hermitian basis for which γ is a multiple of the unit matrix, and g_{abc} is completely antisymmetrical. Relative to this basis, which still has the freedom of orthogonal transformations, the g matrices are antisymmetrical,

$$g_a^T = - g_a \quad , \tag{6.12}$$

and, being imaginary, are Hermitian matrices. Thus the g matrices qualify as a finite-dimensional matrix representation provided they are linearly independent. A linear relation among the g matrices will occur only if a linear combination of the operators $G_{(a)}$, $a = 1,\ldots,\nu$, commutes with every G. Should such a linear combination exist, let it be labelled $G_{(\nu)}$ by an appropriate orthogonal basis transformation. Then

$$[G_{(\nu)} , G_{(b)}] = \sum G_{(a)} \, g_{a\nu b} = 0 \tag{6.13}$$

and $g_{\nu ab} = 0$, which states that $G_{(\nu)}$ will never appear in the expression for any commutator. This procedure can be continued if there are several such linear combinations and we reach the conclusion that the group can be factored into an Abelian group, and a non-commutative group with its structure characterized by the property that the g matrices constitute a finite dimensional representation of the generating operators.

6.3 SUBGROUPS

Groups of the latter type are necessarily semi-simple, by which is meant that they possess no Abelian invariant subgroups (a simple group has no invariant subgroup). The significance of these terms can be given within the framework of infinitesimal transformations. Let the generators be divided into two sets, designated as 1 and 2 , of which the first refers to the subgroup. Then, as the condition for the formation of commutators to be closed within the subgroup, we have

$$\text{subgroup:} \quad g_{a_2 b_1 c_1} = 0 \quad . \tag{6.14}$$

The subgroup is invariant if the commutator of any subgroup element with an outside operator is still within the subgroup,

$$\text{invariant subgroup:} \quad g_{a_2 b_1 c_2} = 0 \quad ; \tag{6.15}$$

and, if the subgroup is Abelian,

$$\text{Abelian subgroup:} \quad g_{a_1 b_1 c_1} = 0 \quad . \tag{6.16}$$

Then, if the group possesses an Abelian invariant subgroup, the only non-zero elements of a matrix g_{b_1} are of the form $g_{a_1 b_1 c_2}$, and the matrix g_{b_1} cannot be antisymmetrical. Alternatively, we conclude from these attributes of an Abelian invariant subgroup that

$$\mathrm{tr}\,(g_{b_1}\, g_{c_1}) = \sum_{ad} g_{ab_1d}\, g_{dc_1a} = 0 \quad , \qquad (6.17)$$

in contradiction with the positive-definiteness this array of numbers should exhibit if the matrices g_{a_1} constitute a finite-dimensional representation. Evidently a group that contains an Abelian invariant subgroup cannot possess finite dimensional matrix representations. A fundamental example of this situation is provided by the group of translations and rotations in three-dimensional space. On referring to the commutation properties (2.80), we recognize that translations form an Abelian invariant subgroup, and the mathematical impossibility of a finite dimensional representation corresponds to the physical existence of an infinite number of states that are connected by the operation of translation. In contrast, the subgroup of rotations, considered by itself, is a simple group and every matrix representation, labelled by the value of the total angular momentum, is of finite dimensionality.

6.4 DIFFERENTIAL FORMS AND COMPOSITION PROPERTIES

The transformation described by the infinitesimal changes $d\tau$ of the parameters is produced by the operator

$$1 + i \sum G_{(r)}\{x(\tau)\tau\}\, d\tau_r = 1 + i \sum G_{(a)}\{x(\tau)\}\, \delta_\ell \tau_a \tag{6.18}$$

where the quantities

$$\delta_\ell \tau_a = \sum \xi_{ar}(\tau)\, d\tau_r \tag{6.19}$$

form a set of inexact differentials (Pfaffians). The subscript ℓ(eft) refers to the manner in which this operator is combined with the operator $U(\tau)$ that produces the finite transformation from the standard zero values of the parameters, namely

$$\begin{aligned} U(\tau+d\tau) &= U(\tau)[1 + i \sum G_{(a)}\{x(\tau)\}\, \delta_\ell \tau_a] \\ &= [1 + i \sum G_{(a)}(x)\, \delta_\ell \tau_a\,]U(\tau) \quad , \end{aligned} \tag{6.20}$$

when the dynamical variables are referred to the standard values of the parameters with the aid of the transformation

$$x(\tau) = U(\tau)^{-1} x U(\tau) \quad . \tag{6.21}$$

This illustrates the general composition property of the group,

$$\begin{aligned} U(\tau) &= U(\tau_1)U(\tau_2) \\ \tau &= \tau(\tau_1 , \tau_2) \quad , \end{aligned} \tag{6.22}$$

for infinitesimal τ_1 . In addition to the infinitesimal transformation $U(\tau+d\tau)\ U(\tau)^{-1}$ one can consider $U(\tau)^{-1}\ U(\tau+d\tau)$, and there must exist a second set of inexact differentials, $\delta_r\tau_a$ such that

$$U(\tau+d\tau) = U(\tau)[1 + i \sum G_{(a)}(x)\ \delta_r\tau_a\ ] \quad . \tag{6.23}$$

An infinitesimal change of τ_1 in the general multiplication property induces a corresponding change of τ and,

$$\begin{aligned} &[1 + i \sum G_{(a)}\ \delta_\ell\tau_a(\tau)\ ]U(\tau) \\ &\mp [1 + i \sum G_{(a)}\ \delta_\ell\tau_a(\tau_1)\ ]\ U(\tau_1)U(\tau_2) \end{aligned} \tag{6.24}$$

or

$$\delta_\ell \tau_a(\tau) = \delta_\ell \tau_a(\tau_1) \quad . \tag{6.25}$$

The ensuing differential equations (Maurer-Cartan)

$$\sum \xi_{ar}(\tau)\, d\tau_r = \sum \xi_{ar}(\tau_1)\, d\tau_{1r} \quad , \tag{6.26}$$

together with the initial conditions

$$\tau_1 = 0 \quad , \quad \tau = \tau_2 \quad , \tag{6.27}$$

serve to determine the composition properties of the group parameters. The same function, as performed by the second set of differentials, is expressed by the differential equations

$$\delta_r\, \tau_a(\tau) = \delta_r\, \tau_a(\tau_2) \tag{6.28}$$

and the initial conditions

$$\tau_2 = 0 \quad , \quad \tau = \tau_1 \quad . \tag{6.29}$$

6.5 CANONICAL PARAMETERS

The choice of parameters is arbitrary to within non-singular transformations, $\tau \to \tau'$, which do not affect the inexact differentials,

$$\delta_\ell \tau_a = \delta_\ell \tau'_a \quad , \tag{6.30}$$

and thus

$$\xi_{as}(\tau') = \sum \xi_{ar}(\tau) \frac{\partial \tau_r}{\partial \tau'_s} \quad . \tag{6.31}$$

The differential equations (6.3) maintain their form under parameter transformation. Through the freedom of parameter and basis changes one could require identity of the basis operators $G_{(a)}$ with the generators $G_{(r)}(\tau)$, for $\tau = 0$. This would be expressed by adding the initial condition

$$\xi_{ab}(0) = \delta_{ab} \tag{6.32}$$

to the ξ differential equations. A special set of parameters, termed canonical, is defined as follows. As the number λ varies from 0 to 1 , let a point in the τ-parameter space move out from the origin along the curve described by

$$\sum \xi_{ar}(\tau)\, d\tau_r = t_a\, d\lambda \quad , \tag{6.33}$$

where the t_a are arbitrary constants. The point in the τ-space that is reached at $\lambda = 1$ is determined by the numbers t_a, which constitute the new set of parameters. According to the invariance of the differential forms the same path is described in the t-parameter space by

$$\sum \xi_{ab}\big(t(\lambda)\big)\ dt_b(\lambda) = t_a\ d\lambda \quad , \qquad t(0) = 0 \quad , \quad t(1) = t \quad . \tag{6.34}$$

Now if t_a is replaced by γt_a, $\gamma < 1$, the point reached for $\lambda = 1$ is identical with the point attained at $\lambda = \gamma$ along the curve characterizing the point t. Hence the path appears in the t-space as

$$t_a(\lambda) = \lambda t_a \quad , \tag{6.35}$$

a straight line, and (6.34) asserts that

$$\delta t_a = dt_a \quad , \tag{6.36}$$

or

$$\sum \xi_{ab}(t)\ t_b = t_a \quad . \tag{6.37}$$

The operator producing the infinitesimal transformation characterized by $d\lambda$ is

$$1 + i\left[\sum G_{(a)} t_a\right] d\lambda \quad , \tag{6.38}$$

and thus the finite transformations of the group have the exponential form

$$U(t) = e^{i \sum G_{(a)} t_a} \quad , \tag{6.39}$$

in terms of the canonical parameters.

The differential equations for the functions $\xi_{ab}(t)$ can be simplified with the aid of the property (6.37). Indeed,

$$\sum_c t_c\left(\frac{\partial}{\partial t_c} \xi_{ab} - \frac{\partial}{\partial t_b} \xi_{ac}\right) = i \sum_{cde} g_{ade}\, \xi_{dc}\, t_c\, \xi_{eb} \tag{6.40}$$

becomes

$$\left(\sum_c t_c \frac{\partial}{\partial t_c} + 1\right) \xi_{ab} - i \sum_{dc} g_{ade}\, t_d\, \xi_{eb} = \delta_{ab} \tag{6.41}$$

and, on applying this equation at the point λt , we obtain

$$\frac{\partial}{\partial\lambda}\lambda\xi_{ab}(\lambda t) - i\sum g_{ade}\, t_d\, \lambda\, \xi_{eb}(\lambda t) = \delta_{ab} \quad . \tag{6.42}$$

In a matrix notation the latter reads

$$\left(\frac{\partial}{\partial\lambda} - i\sum g_a t_a\right)\lambda\xi(\lambda t) = 1 \quad , \tag{6.43}$$

and the formal solution is

$$\xi(t) = \int_0^1 d\lambda\, e^{i\lambda\sum g_a t_a} = \frac{e^{i\sum g_a t_a} - 1}{i\sum g_a t_a} \quad . \tag{6.44}$$

Thus the differential forms $\delta_\ell t$ are obtained explicitly as

$$\delta_\ell t = \frac{e^{igt} - 1}{igt}\, dt \quad , \tag{6.45}$$

in which we have used an evident notation. The structure of $\delta_r t$ can now be inferred from the following property of the canonical parameters,

$$U(t)^{-1} = U(-t) \quad . \tag{6.46}$$

The substitution $t \to -t-dt$, $dt \to dt$, converts $U(t+dt)\, U(t)^{-1}$ into $U(t)^{-1}\, U(t+dt)$ and therefore

$$\delta_r t = \xi(-t)\, dt = \frac{1 - e^{-igt}}{igt}\, dt \quad . \tag{6.47}$$

The general expression of this connection between the two differential forms, characteristic of the canonical parameters, is the group composition property

$$t = t(t_1 \, , \, t_2) = - \, t(-t_2 \, , \, -t_1) \quad . \tag{6.48}$$

6.6 AN EXAMPLE. SPECIAL CANONICAL GROUP

A simple illustration of these considerations is provided by a group of three parameters defined by the commutation relations

$$\begin{aligned} [G_{(1)} \, , \, G_{(2)}] &= iG_{(3)} \quad , \\ [G_{(1)} \, , \, G_{(3)}] &= [G_{(2)} \, , \, G_{(3)}] = 0 \quad . \end{aligned} \tag{6.49}$$

The matrices g_a are conveniently presented in the linear combination

$$igt = \begin{pmatrix} 0 & 0 & 0 \\ 0 & 0 & 0 \\ t_2 & -t_1 & 0 \end{pmatrix} \quad , \tag{6.50}$$

which asymmetrical form shows that

$$(igt)^2 = 0 \quad . \tag{6.51}$$

Thus the matrices g_a do not furnish a finite-dimensional Hermitian matrix representation, which is related to the existence of the one-parameter invariant subgroup generated by $G_{(3)}$. In virtue of the algebraic property (6.51), we have

$$\xi(t) = 1 + \tfrac{1}{2}igt \quad , \tag{6.52}$$

and the differential equations (6.26) read (primes are now used to distinguish the various parameter sets)

$$\begin{gathered} dt_1 = dt_1' \quad , \quad dt_2 = dt_2' \quad , \\ dt_3 + \tfrac{1}{2}(t_2\, dt_1 - t_1\, dt_2) \\ = dt_3' + \tfrac{1}{2}(t_2'\, dt_1' - t_1'\, dt_2') \quad . \end{gathered} \tag{6.53}$$

The solution of these equations subject to the initial condition $t' = 0 \ : \ t = t''$, and the canonical parameter composition law of the group, is

$$t_1 = t_1' + t_1'' \quad , \quad t_2 = t_2' + t_2''$$
$$t_3 = t_3' + t_3'' - \tfrac{1}{2}\left(t_1' t_2'' - t_2' t_1''\right) \quad , \tag{6.54}$$

which illustrates the reflection property (6.48). The operator statement contained here is

$$e^{iGt'} e^{iGt''} = e^{iG(t'+t'') - \frac{1}{2}iG_{(3)}\left(t_1' t_2'' - t_2' t_1''\right)} \tag{6.55}$$

and specializations of this result, in which the parameters are combined with the generators, can be presented as

$$e^{i\left(G_1+G_2\right)} = e^{iG_1} e^{iG_2} e^{\frac{1}{2}[G_1 , G_2]}$$
$$= e^{iG_2} e^{iG_1} e^{-\frac{1}{2}[G_1 , G_2]} \quad . \tag{6.56}$$

It will not have escaped attention that the commutation properties (6.49) are realized by the special canonical group. Accordingly, the operators expressing the finite transformations of this group

$$U(q'p'\lambda') = e^{i(pq'-p'q+\lambda')} \quad , \tag{6.57}$$

possess the multiplication property

$$U(q'p'\lambda')U(q''p''\lambda'') = U\left(q'+q'' \;,\; p'+p'' \;,\; \lambda'+\lambda'' + \tfrac{1}{2}(p'q''-p''q')\right) \tag{6.58}$$

which is applicable to all types of dynamical variables. Specializations analogous to (6.56) are

$$\begin{aligned} e^{i(pq'-p'q)} &= e^{ipq'}\, e^{-ip'q}\, e^{\frac{1}{2}ip'q'} \\ &= e^{-ip'q}\, e^{ipq'}\, e^{-\frac{1}{2}ip'q'} \quad . \end{aligned} \tag{6.59}$$

6.7 OTHER PARAMETERS. ROTATION GROUP

Canonical parameters are not always the most useful parameter choice. This can be illustrated by the three-dimensional rotation group. We combine the three canonical parameters into the vector ω , and observe that the three-dimensional Hermitian matrices $j(=g_a)$, defined by the commutation relations (2.81), can also be presented as a vector operation:

$$i\omega\cdot j = \begin{pmatrix} 0 & \omega_3 & -\omega_2 \\ -\omega_3 & 0 & \omega_1 \\ \omega_2 & -\omega_1 & 0 \end{pmatrix} = -\,\omega\times \quad . \tag{6.60}$$

Hence

$$(i\omega\cdot\mathbf{j})^2 = \omega\times(\omega\times = \omega\omega\cdot - \omega^2 \quad , \tag{6.61}$$

and

$$(i\omega\cdot\mathbf{j})^3 = -\omega^2(i\omega\cdot\mathbf{j}) \quad . \tag{6.62}$$

The latter result shows that the eigenvalues of any component of $\mathbf{j}$ are 1, 0, -1, and therefore

$$f(\omega\cdot\mathbf{j}) = \left[1 - \left(\frac{\omega\cdot\mathbf{j}}{\omega}\right)^2\right] f(0) + \frac{1}{2}\frac{\omega\cdot\mathbf{j}}{\omega}\left(\frac{\omega\cdot\mathbf{j}}{\omega} + 1\right) f(\omega)$$
$$+ \frac{1}{2}\frac{\omega\cdot\mathbf{j}}{\omega}\left(\frac{\omega\cdot\mathbf{j}}{\omega} - 1\right) f(-\omega) \quad . \tag{6.63}$$

On applying this result to the explicit construction of $\delta_\ell\omega$, as given in (6.45), we obtain

$$\delta_\ell\omega = \left[1 + i\,\frac{\omega\cdot\mathbf{j}}{\omega}\,\frac{1-\cos\omega}{\omega} - \left(\frac{\omega\cdot\mathbf{j}}{\omega}\right)^2\left(1 - \frac{\sin\omega}{\omega}\right)\right] d\omega$$
$$= \frac{\sin\omega}{\omega}\,d\omega$$
$$+ \left(1 - \frac{\sin\omega}{\omega}\right)\frac{\omega\;\omega\cdot d\omega}{\omega^2} - \frac{1-\cos\omega}{\omega^2}\,\omega\times d\omega \quad , \tag{6.64}$$

or, with a simple rearrangement,

$$\tfrac{1}{2}\,\delta_\ell \boldsymbol{\omega} = \cos \tfrac{1}{2}\omega \; d\left(\frac{\sin \tfrac{1}{2}\omega}{\omega}\, \boldsymbol{\omega} \right) - \frac{\sin \tfrac{1}{2}\omega}{\omega}\, \boldsymbol{\omega}\; d(\cos \tfrac{1}{2}\omega)$$

$$- \frac{\sin \tfrac{1}{2}\omega}{\omega}\, \boldsymbol{\omega} \times d\left(\frac{\sin \tfrac{1}{2}\omega}{\omega}\, \boldsymbol{\omega} \right) \quad . \tag{6.65}$$

Thus, the new parameters

$$u_0 = \cos \tfrac{1}{2}\omega \quad , \quad \mathbf{u} = \frac{\sin \tfrac{1}{2}\omega}{\omega}\, \boldsymbol{\omega} \quad , \tag{6.66}$$

$$u_0^2 + \mathbf{u}^2 = 1 \quad , \tag{6.67}$$

are such that

$$\tfrac{1}{2}\,\delta_\ell \boldsymbol{\omega} = u_0\, d\mathbf{u} - \mathbf{u}\, du_0 - \mathbf{u} \times d\mathbf{u} \quad . \tag{6.68}$$

The substitution $\boldsymbol{\omega} \to -\boldsymbol{\omega}$ converts $\mathbf{u}$ into $-\mathbf{u}$ while leaving u_0 unaltered, and therefore the analogous expression for the differential form $\delta_r\boldsymbol{\omega}$ is

$$\tfrac{1}{2}\,\delta_r \boldsymbol{\omega} = u_0\, d\mathbf{u} - \mathbf{u}\, du_0 + \mathbf{u} \times d\mathbf{u} \quad . \tag{6.69}$$

Although u_0 is not an independent parameter, it can largely be treated as such. This is indicated by the structure of the differentials $d\mathbf{u}$, du_0 that a given $\delta_\ell\boldsymbol{\omega}$ or $\delta_r\boldsymbol{\omega}$ imply:

$$\begin{aligned} d\mathbf{u} &= u_0 \tfrac{1}{2}\, \delta_\ell \omega + \mathbf{u}\times\tfrac{1}{2}\, \delta_\ell \omega \\ du_0 &= - \mathbf{u}\cdot\tfrac{1}{2}\, \delta_\ell \omega \quad , \end{aligned} \tag{6.70}$$

and

$$\begin{aligned} d\mathbf{u} &= u_0 \tfrac{1}{2}\, \delta_r \omega - \mathbf{u}\times\tfrac{1}{2}\, \delta_r \omega \\ du_0 &= - \mathbf{u}\cdot\tfrac{1}{2}\, \delta_r \omega \quad , \end{aligned} \tag{6.71}$$

for these changes maintain the normalization (6.67),

$$d\left(u_0^2 + \mathbf{u}^2\right) = 0 \quad . \tag{6.72}$$

Thus, there is a unit vector in a four-dimensional Euclidean space associated with every three-dimensional rotation, and, to the composition of two three-dimensional rotations there is associated a four-dimensional rotation. The algebraic simplification achieved by the u-parameters appears in the group composition law. The invariance of the bilinear form for the differentials $\delta_\ell \omega$, expressed by the differential equations (6.26), implies a linear relation between the parameters of the individual transformations and those of the product transformation. One easily verifies that

$$\mathbf{u} = u_0'\mathbf{u}'' + \mathbf{u}'u_0'' - \mathbf{u}'\times\mathbf{u}''$$
$$u_0 = u_0'u_0'' - \mathbf{u}'\cdot\mathbf{u}'' \quad . \qquad (6.73)$$

The u-parameters appear in another way on recalling the existence of automorphisms - unitary transformations - of the algebra defined by 2n Hermitian canonical variables of the second kind, ξ_κ , that have the structure of the Euclidean rotation group in $2n + 1$ dimensions. Hence, a representation of the three-dimensional rotation group is generated by such unitary transformations of the three anticommutative Hermitian operators ξ_1 , ξ_2 , $\xi_3 = -i2^{\frac{1}{2}}\xi_1\xi_2$, or, of the operators

$$\sigma_k = 2^{\frac{1}{2}}\xi_k \quad , \quad k = 1,\ldots,3 \quad , \qquad (6.74)$$

which have the following multiplication characteristics:

$$\sigma_k\sigma_\ell = \delta_{k\ell} + i\sum_m \varepsilon_{k\ell m}\,\sigma_m \quad , \qquad (6.75)$$

where ε is the completely antisymmetrical function of its indices specified by $\varepsilon_{123} = +1$. The generators of the three independent infinitesimal rotations are the operators $\frac{1}{2}\sigma$ for, according to (4.96),

$$\xi_{k\ell} = \frac{1}{4i} [\sigma_k , \sigma_\ell] = \sum_m \varepsilon_{k\ell m} \tfrac{1}{2}\sigma_m \qquad (6.76)$$

and

$$\xi_{12} = \tfrac{1}{2}\sigma_3 \quad , \qquad (6.77)$$

for example. Thus, the simplest non-trivial measurement algebra, of dimensionality 2^2 , provides an angular momentum operator matrix representation

$$\tfrac{1}{2}\sigma \times \tfrac{1}{2}\sigma = i\tfrac{1}{2}\sigma \quad . \qquad (6.78)$$

Explicit matrix representations are obtained on relating the operator basis $1 , \sigma$ to measurement symbols. The measurement symbols of the σ_3 representation are presented in the following array,

$$M = \begin{pmatrix} \frac{1}{2}(1+\sigma_3) & \frac{1}{2}(\sigma_1+i\sigma_2) \\ \frac{1}{2}(\sigma_1-i\sigma_2) & \frac{1}{2}(1-\sigma_3) \end{pmatrix} \qquad (6.79)$$

and (Pauli)

$$\sigma_1 = \begin{pmatrix} 0 & 1 \\ 1 & 0 \end{pmatrix} \quad , \quad \sigma_2 = \begin{pmatrix} 0 & -i \\ i & 0 \end{pmatrix} \quad , \quad \sigma_3 = \begin{pmatrix} 1 & 0 \\ 0 & -1 \end{pmatrix} \quad . \tag{6.80}$$

Now, apart from the freedom of multiplication by a numerical phase factor, any unitary operator of this algebra has the form

$$\begin{aligned} \mathcal{U} &= u_0 + i\mathbf{u}\cdot\boldsymbol{\sigma} \\ \mathcal{U}^\dagger &= \mathcal{U}^{-1} = u_0 - i\mathbf{u}\cdot\boldsymbol{\sigma} \quad , \end{aligned} \tag{6.81}$$

where the four numbers comprised in u_0 , $\mathbf{u}$ are real, and obey

$$u_0^2 + \mathbf{u}^2 = 1 \quad . \tag{6.82}$$

The latter condition also states that $\mathcal{U}$ is unimodular,

$$\det \mathcal{U} = +1 \quad . \tag{6.83}$$

With any such unitary operator there is associated a three-dimensional proper rotation

$$\begin{aligned} \mathcal{U}^{-1}\sigma_k\mathcal{U} &= \sum_\ell r_{k\ell}\,\sigma_\ell \\ r\,r^T = 1 \quad &, \quad \det r = +1 \quad , \end{aligned} \tag{6.84}$$

which correspondence is 2:1 since U and $-U$ produces the same rotation, and successive unitary transformations generate successive rotations. The parameter composition law that emerges from the product

$$u_0 + i\mathbf{u}\cdot\sigma = \left(u_0' + i\mathbf{u}'\cdot\sigma\right)\left(u_0'' + i\mathbf{u}''\cdot\sigma\right) \qquad (6.85)$$

is just (6.73). Incidentally, the explicit form of the three-dimensional rotation matrix r is

$$r = \frac{u_0 + i\mathbf{u}\cdot\mathbf{j}}{u_0 - i\mathbf{u}\cdot\mathbf{j}} = 1 + 2iu_0\mathbf{u}\cdot\mathbf{J} - 2(\mathbf{u}\cdot\mathbf{j})^2 \qquad (6.86)$$

where $\mathbf{j}$ signifies the three-dimensional angular momentum matrix representation displayed in (6.60).

6.8 DIFFERENTIAL OPERATOR REALIZATIONS

The differentiable manifold of the group parameters enables differential operator realizations of the infinitesimal generators of a group to be constructed. Let us define for this purpose two sets of functions $\eta_{as}(\tau)$, $\zeta_{as}(\tau)$, according to

$$d\tau = \delta_\ell \tau \; \eta(\tau) = \delta_r \tau \; \zeta(\tau) \quad . \tag{6.87}$$

Thus,

$$\xi(\tau)^T \eta(\tau) = 1 \quad , \tag{6.88}$$

and, using canonical parameters

$$\zeta(t) = \eta(-t) \quad . \tag{6.89}$$

Now, the infinitesimal composition properties stated in (6.20) and (6.23) can be presented as

$$dU = i \; \delta_\ell \tau \; GU = i \; \delta_r \tau \; UG \quad , \tag{6.90}$$

or, on writing

$$\begin{aligned} dU = d\tau \frac{\partial}{\partial \tau} U &= \delta_\ell \tau \; \eta(\tau) \frac{\partial}{\partial \tau} U \\ &= \delta_r \tau \; \zeta(\tau) \frac{\partial}{\partial \tau} U \quad , \end{aligned} \tag{6.91}$$

as

$$\begin{aligned} - GU = - \eta(\tau) \frac{1}{i} \frac{\partial}{\partial \tau} U &= \mathcal{G}_\ell U \\ UG = \zeta(\tau) \frac{1}{i} \frac{\partial}{\partial \tau} U &= \mathcal{G}_r U \quad . \end{aligned} \tag{6.92}$$

The two sets of differential operators defined here are commutative,

$$\mathcal{G}_{(a)\ell}\ \mathcal{G}_{(b)r}\ U = \mathcal{G}_{(b)r}\ \mathcal{G}_{(a)\ell}\ U = -\ G_{(a)}\ U\ G_{(b)} \quad , \tag{6.93}$$

and each set obeys the G commutation relations

$$\begin{aligned} &[\mathcal{G}_{(b)\ell}\ ,\ \mathcal{G}_{(c)\ell}]U = -\ [G_{(b)}\ ,\ G_{(c)}]U \\ &= -\sum_a G_{(a)}\ g_{abc}\ U = \sum_a \mathcal{G}_{(a)\ell}\ g_{abc}\ U \quad , \\ &[\mathcal{G}_{(b)r}\ ,\ \mathcal{G}_{(c)r}]U = U[G_{(b)}\ ,\ G_{(c)}] \\ &= U\sum_a G_{(a)}\ g_{abc} = \sum_a \mathcal{G}_{(a)r}\ g_{abc}\ U \quad . \end{aligned} \tag{6.94}$$

These are intrinsic properties of the differential operators. Thus, one can verify that the differential equations implied for $\eta(\tau)$,

$$\sum_r \left(\eta_{br}\ \frac{\partial}{\partial \tau_r} \eta_{cs} - \eta_{cr}\ \frac{\partial}{\partial \tau_r}\ \eta_{bs}\right) = -\ i \sum_a \eta_{as}\ g_{abc} \tag{6.95}$$

are a direct consequence of (6.3) and the relation (6.88).

6.9 GROUP VOLUME

An infinitesimal element of volume can be defined on the group manifold with the aid of the inexact differentials (6.19)

$$d[\tau] = |\det \xi(\tau)| \; (d\tau) \; . \qquad (6.69)$$

This volume element is independent of the choice of parameters, and is unchanged by the parameter transformation, $\tau_1 \rightarrow \tau$, that expresses group multiplication, $\tau = \tau(\tau_1 \, , \, \tau_2)$. An alternative definition of volume accompanies the differentials $\delta_r\tau$, and that volume element is invariant under the parameter transformation $\tau_2 \rightarrow \tau$ of group multiplication. The relation

$$\xi(-t) = e^{-igt} \, \xi(t) \quad , \qquad (6.97)$$

and its consequence

$$\det \xi(-t) = e^{-i \operatorname{tr} gt} \det \xi(t) \quad , \qquad (6.98)$$

shows that the two definitions of volume are identical if

$$\operatorname{tr} g_b = \sum_a g_{aba} = 0 \; . \qquad (6.99)$$

The latter property certainly holds if the group possesses a finite-dimensional matrix representation, for then all the g matrices are antisymmetrical (or zero) to within the latitude of matrix transformations that do not alter the trace. But, as the example of (6.50) indicates, this is by no means a necessary condition. The invariance of the volume element, stated by

$$\det \xi(\tau)\ (d\tau) = \det \xi(\tau_1)\ (d\tau_1) \quad , \tag{6.100}$$

is given a differential form on choosing the parameters τ_2 to be infinitesimal, in which circumstance the explicit transformation is

$$\tau = \tau_1 + \tau_2 \zeta(\tau_1) \quad , \tag{6.101}$$

and we infer that

$$\sum_r \frac{\partial}{\partial \tau_r} [\zeta_{ar}(\tau) \det \xi(\tau)] = 0 \ . \tag{6.102}$$

As an application of this result, let us observe that

$$(\det \xi)^{\frac{1}{2}} \mathcal{G}_{(a)r} (\det \xi)^{-\frac{1}{2}}$$

$$= \sum_r \zeta_{as} \frac{1}{i} \frac{\partial}{\partial \tau_s} - \tfrac{1}{2} \sum_s \zeta_{as} \frac{1}{i} \left(\frac{\partial}{\partial \tau_s} \log \det \xi \right) \tag{6.103}$$

$$= \sum_r \tfrac{1}{2} \left\{ \zeta_{as}(\tau) \ , \frac{1}{i} \frac{\partial}{\partial \tau_s} \right\} \quad ,$$

which transformation maintains the commutation properties of the differential operators, and yields a formally Hermitian differential operator to represent a Hermitian generating operator $G_{(a)}$. The study of a ν-parameter group of unitary transformations can thus be performed with the aid of an equivalent dynamical system described by ν pairs of complementary variables of the first kind which are generally quasi-canonical for, unless the range of the individual parameters is $-\infty$ to ∞ , these variables do not possess all the attributes of canonical variables.

6.10 COMPACT GROUPS

The ability to integrate over the group manifold is particularly valuable when the group is compact, which is to say that any infinite sequence of group elements possesses a limit point belonging to the group manifold. Thus the manifold of a compact group is bounded, and its volume can be chosen as unity by including a suitable scale factor in the volume element. We first notice that the matrices g_a for a compact group are necessarily traceless and, accordingly, the two definitions of volume element are identical. To prove this consider a particular transformation U, as characterized by canonical parameters t, and the corresponding finite matrix

$$\mathcal{U}(t) = e^{igt} \tag{6.104}$$

which appears in the general relation

$$U\,G\,U^{-1} = G\mathcal{U}\ , \tag{6.105}$$

of which (6.5) is the infinitesimal transformation form. If the imaginary matrices g are not antisymmetrical,

$$\det \mathcal{U}(t) = e^{i\,\mathrm{tr}\,gt} \tag{6.106}$$

may differ from unity. Then, to the sequence of operators U^k, $k = \pm 1\ , \pm 2\ , \ldots$, there corresponds a sequence of matrices $\mathcal{U}^k$, for which

$$\det \mathcal{U}^k = (\det \mathcal{U})^k \tag{6.107}$$

increases without limit as $k \to +\infty$, if $\det \mathcal{U} > 1$, or, as $k \to -\infty$, if $\det \mathcal{U} < 1$. This contradicts the requirement that an infinite sequence of group elements possess a limit point on the group manifold, with its associated finite $\mathcal{U}$ matrix. Hence the matrix $\mathcal{U}$ must be unimodular, and every g_a has a vanishing trace. It may be noted here that a group for which the g matrices supply a representation of its infinitesimal Hermitian generators has a bounded manifold. According to the explicit construction of the volume element in terms of the canonical parameters,

$$d[t] = C \left| \det \left(\frac{\sin \frac{1}{2}gt}{\frac{1}{2}gt} \right) \right| (dt) \quad , \qquad (6.108)$$

and the boundaries of the manifold are reached when the weight factor in the element of volume vanishes. The statement that the g matrices are Hermitian and linearly independent implies that, for every t, gt possesses non-zero, real eigenvalues. The numerically largest of these eigenvalues equated to 2π, then determines the finite points where the boundary of the group manifold intercepts the ray, directed from the origin of the parameter space, which is specified by the relative values of the parameters t_a.

6.11 PROJECTION OPERATORS AND INVARIANTS

The group property and the invariance aspects of the volume element for a compact group assert that the operator

$$P_0 = \int d[\tau]\ U(\tau) \tag{6.109}$$

has the following characteristics

$$\begin{aligned} P_0 U(\tau_2) &= \int d[\tau_1]\ U(\tau_1) U(\tau_2) = \int d[\tau]\ U(\tau) \\ &= P_0 = U(\tau_1) P_0 \quad , \\ P_0^2 = P_0 \int d[\tau_2]\ U(\tau_2) &= P_0 \int d[\tau_2] \\ &= P_0 \quad , \end{aligned} \tag{6.110}$$

and, using the canonical parameters,

$$P_0^\dagger = \int d[t]\ U(-t) = P_0 \quad , \tag{6.111}$$

since intergration with respect to t, or $-t$, covers the group manifold and the volume element is invariant under the transformation $t \to -t$. Thus, the Hermitian operator P_0 is a measurement symbol or, in geometrical language, a projection operator, for the subspace of states that are invariant under all the transformations of the group. These are also the states for which all the generating operators can be simultaneously assigned the value zero. In a similar way

$$U(\tau_2)^{-1} \left(\int d[\tau_1]\ U(\tau_1)^{-1} X U(\tau_1) \right) U(\tau_2)$$

$$= \int d[\tau]\ U(\tau)^{-1} X U(\tau) \tag{6.112}$$

describes the construction of the subalgebra of operators that are invariant under all transformations of the group. One can also apply a slight modification of the latter procedure to the finite, real matrices $\mathcal{U}(\tau)$, with the result

$$\mathcal{U}(\tau_2)^T \left(\int d[\tau_1]\ \mathcal{U}(\tau_1)^T \mathcal{U}(\tau_1) \right) \mathcal{U}(\tau_2)$$

$$= \int d[\tau]\ \mathcal{U}(\tau)^T \mathcal{U}(\tau) \quad . \tag{6.113}$$

This is a real symmetrical positive definite matrix and therefore it can be expressed as the square of a matrix of the same type, say λ . Thus the content of (6.113) is

$$\left(\lambda \mathcal{U}(\tau)\ \lambda^{-1}\right)^T \left(\lambda\ \mathcal{U}(\tau)\ \lambda^{-1}\right) = 1 \quad , \tag{6.114}$$

whish is to say that a basis for the Hermitian generating operators of a compact group can be found that implies real orthogonal, or unitary, matrices $\mathcal{U}(\tau)$, and antisymmetrical matrices g_a .

If the operator X in (6.112) is chosen as an algebraic function of the generating operators G , the integration process will produce those algebraic functions that commute with every G and therefore serve to classify, by their eigenvalues the various matrix representations of the infinitesimal generators of the group. Now

$$\int d[\tau]\; U(\tau)^{-1} f(G) U(\tau) = \int d[\tau]\; f\big(U(\tau)^{-1} G U(\tau)\big)$$
$$= \int d[\tau]\; f\big(\mathcal{U}(\tau) G\big) \quad , \qquad (6.115)$$

in which it is supposed that the basis is suitable to produce a unitary matrix $\mathcal{U}$. The effect of the integration is achieved by requiring that

$$f\big(\mathcal{U}(\tau) G\big) = f(G) \qquad (6.116)$$

and, since f(G) can be chosen as a symmetrical homogeneous function of the various G_a , the operator nature of the latter is not relevant which permits (6.116) to be replaced by the numerical invariance requirement

$$f\big(\mathcal{U}(\tau)\gamma\big) = f(\gamma) \quad , \qquad (6.117)$$

referring to functions of a vector γ in a ν-dimensional space. The infinitesimal version of this invariance property can be expressed as

$$\mathcal{G} f(\gamma) = 0 \tag{6.118}$$

where the differential operators

$$\mathcal{G}_b = \gamma g_b \frac{\partial}{\partial \gamma} = \sum_{ac} \gamma_a \, g_{abc} \frac{\partial}{\partial \gamma_c} \tag{6.119}$$

are realizations of the generating operators (for a semi-simple group),

$$[\mathcal{G}_a \, , \mathcal{G}_b] = \gamma [g_a \, , g_b] \frac{\partial}{\partial \gamma} \quad , \tag{6.120}$$

which are analogous to, but are less general than the differential operators $\mathcal{G}_\ell$ and $\mathcal{G}_r$. A set of invariant functions can be constructed directly if a finite dimensional matrix representation is known. Let **G** and **U** symbolize a κ-dimensional matrix representation of the corresponding Hermitian and unitary operators. Then

$$\mathbf{U}(\tau)(\lambda - \mathbf{G}\gamma) \, \mathbf{U}(\tau)^{-1} = \lambda - \mathbf{G} \mathcal{U}(\tau)\gamma \quad , \tag{6.121}$$

and

$$f(\gamma) = \det(\lambda - \mathbf{G}\gamma)$$

$$= \lambda^{\kappa} - \lambda^{\kappa-1}\,\mathrm{tr}\,\mathbf{G}\gamma - \lambda^{\kappa-2}\tfrac{1}{2}\left(\mathrm{tr}\,(\mathbf{G}\gamma)^2 - (\mathrm{tr}\,\mathbf{G}\gamma)^2\right) + \ldots \tag{6.122}$$

is an invariant function. The coefficients of the powers of λ, or equivalently, the traces

$$\mathrm{tr}\left(\sum_a \mathbf{G}_{(a)}\gamma_a\right)^k , \qquad k = 1,\ldots,\kappa \tag{6.123}$$

supply invariant symmetrical functions of γ, and thereby of the operators G.

6.12 DIFFERENTIAL OPERATORS AND THE ROTATION GROUP

The differential and integral group properties we have been discussing can be illustrated with the three-dimensional rotation group. On referring to (6.70-1) we see that the two sets of differential operators that realize the abstract angular momentum operators are

$$\begin{aligned} \mathcal{J}_\ell &= \frac{1}{2i}\left[\mathbf{u}\times\frac{\partial}{\partial\mathbf{u}} + \mathbf{u}\frac{\partial}{\partial u_0} - u_0\frac{\partial}{\partial\mathbf{u}}\right] \\ \mathcal{J}_r &= \frac{1}{2i}\left[\mathbf{u}\times\frac{\partial}{\partial\mathbf{u}} - \mathbf{u}\frac{\partial}{\partial u_0} + u_0\frac{\partial}{\partial\mathbf{u}}\right] \end{aligned} \tag{6.124}$$

or, if we introduce the notation

$$\mathcal{J}_{ab} = u_a \frac{1}{i} \frac{\partial}{\partial u_b} - u_b \frac{1}{i} \frac{\partial}{\partial u_a} \quad , \qquad a,b = 0,\ldots,3 \quad , \tag{6.125}$$

these differential operators appear as

$$\begin{aligned} \mathcal{J}_{\ell 3} &= \tfrac{1}{2}[\mathcal{J}_{12} + \mathcal{J}_{30}] \\ \mathcal{J}_{r3} &= \tfrac{1}{2}[\mathcal{J}_{12} - \mathcal{J}_{30}] \quad , \end{aligned} \tag{6.126}$$

together with the results of cyclically permuting the indices 123 . The differential operators $\mathcal{J}_{ab}$ are such that

$$\mathcal{J}_{ab}\, f\Big(\sum_0^3 u_c^{\,2} \Big) = 0 \quad , \tag{6.127}$$

and are evidently associated with infinitesimal rotations in a four-dimensional Euclidean space. (They obey the four-dimensional extension of the angular momentum commutation relations given in (2.80).) Of course, the group manifold is three-dimensional and, on restoring u_0 as a function of the independent parameters $\mathbf{u}$, the term containing $\partial/\partial u_0$ is omitted in (6.124). The three-dimensional element of volume contains $(d\mathbf{u}) = du_1\, du_2\, du_3$, together with the factor

$|\det \xi|$ which can be evaluated directly or inferred from the relevant form of the differential equations (6.102),

$$-\frac{\partial}{\partial \mathbf{u}} \times (\mathbf{u} \det \xi) + \frac{\partial}{\partial \mathbf{u}} \left(u_0 \det \xi\right) = 0 \quad . \qquad (6.128)$$

The unique solution of this equation states the constancy of $u_0 \det \xi$. Alternatively, one can employ the four variables $u = u_0$, $\mathbf{u}$, and a volume element proportional to $(du) = du_0\,(d\mathbf{u})$ with the restriction on the variables u_0 enforced by a delta function factor $\delta\left(u_0^2 + \mathbf{u}^2 - 1\right)$. On integrating over u_0, one regains the factor $|u_0|^{-1}$ which, in the previous method, is supplied by $|\det \xi|$. Thus, the group manifold is the three-dimensional surface of a four-dimensional Euclidean unit sphere, and the volume element can be described intrinsically, or in terms of the space in which the manifold is imbedded,

$$d[u] = \frac{1}{2\pi^2} \frac{(d\mathbf{u})}{|u_0|} = \frac{1}{\pi^2} \delta\left(u^2 - 1\right) (du) \quad . \qquad (6.129)$$

Here the proper constants have been supplied to normalize the total volume to unity, although in the three-dimensional form one must also sum over the two pieces of the group that correspond to $u_0 = \pm \left(1-\mathbf{u}^2\right)^{\frac{1}{2}}$. The advantage of the four-dimensional form stems from the commutativity of $\delta(u^2 -1)$ with the differential operators $\mathcal{J}$, which permits the latter to be defined in the unbounded four-dimensional Euclidean space, thereby identifying the four variables u_a with canonical variables of the first kind. Thus the general properties of a three-dimensional angular momentum can be studied in terms of an equivalent system consisting of a particle in a four-dimensional Euclidean space, with the correspondence between its orbital angular momentum and the general three-dimensional angular momentum described by (6.126).

There is only one independent operator that commutes with every component of $\mathbf{J}$, namely, $\mathbf{J}^2$, for there is only one independent rotationally invariant function of a three-dimensional vector. We observe from (6.124-6) that

$$\begin{aligned} \mathcal{J}_\ell{}^2 = \mathcal{J}_r{}^2 &= \tfrac{1}{4} \sum_{a<b} \mathcal{J}_{ab}{}^2 \ , \\ \mathcal{J}_{12}\mathcal{J}_{30} + \mathcal{J}_{23}\mathcal{J}_{10} &+ \mathcal{J}_{31}\mathcal{J}_{20} = 0 \ , \end{aligned} \tag{6.130}$$

where the square of the three-dimensional angular momentum vector is represented by the four-dimensional differential operator

$$\tfrac{1}{4} \sum_{a<b} \mathcal{J}_{ab}{}^2 = \tfrac{1}{2}u \frac{\partial}{\partial u} \left(\tfrac{1}{2}u \frac{\partial}{\partial u} + 1\right) - \tfrac{1}{4}u^2 \sum_a \frac{\partial^2}{\partial u_a^2} \ . \quad (6.131)$$

Thus eigenfunctions of the total angular momentum are obtained from four-dimensional spherical harmonics [2] - solutions of the four-dimensional Laplace equation that are homogeneous of integral degree $n = 0, 1, 2, \ldots,$ - and the eigenvalues are

$$\left(J^2\right)' = j(j+1) \ ,$$
$$j = \tfrac{1}{2}n = 0 \ , \tfrac{1}{2} \ , 1 \ , \ldots \ . \quad (6.132)$$

---

[2] Another procedure uses complex combinations of the parameters, as comprised in the pair of complex numbers that form any row or column of the matrix

$$\mathcal{U} = u_0 + i\mathbf{u}\cdot\sigma = \begin{pmatrix} u_0+iu_3 & iu_1+u_2 \\ iu_1-u_2 & u_0-iu_3 \end{pmatrix} \ .$$

Let $y^{\dagger\prime}$ indicate either two component row vector, with y' similarly designating a column vector. The differential composition properties.

$$d\mathcal{U} = i\tfrac{1}{2}\sigma \cdot \delta\boldsymbol{\omega}_\ell \ \mathcal{U} = \mathcal{U} i\tfrac{1}{2}\sigma \cdot \delta\boldsymbol{\omega}_r$$

imply

The completeness of these eigenfunctions, and of the three-dimensional angular momentum spectrum can be inferred from the structure of the fundamental solution, of degree -2 , which refers to the inhomogeneous equation

$$-\sum_a \frac{\partial^2}{\partial u_a^2} [\sum_b (u_b-u_b')^2]^{-2} = \delta[u-u'] \quad . \qquad (6.133)$$

6.13 NON-COMPACT GROUP INTEGRATION

The technique of group integration can also be effective for groups that are not compact. We shall illustrate this with the special canonical group referring to Hermitian variables of the first kind. As the analoque of the operator appearing in (6.112), we consider

---

$$dy' = i\tfrac{1}{2}\sigma \cdot \delta\omega_\ell \, y' \quad , \quad dy^{\dagger\prime} = y^{\dagger\prime} i\tfrac{1}{2}\sigma \cdot \delta\omega_r \ .$$

The corresponding differential operator realizations of an angular momentum vector are:

$$-\frac{\partial}{\partial y'} \tfrac{1}{2}\sigma y' \quad , \quad y^{\dagger\prime}\tfrac{1}{2}\sigma \frac{\partial}{\partial y^{\dagger\prime}} \ .$$

The squared angular momentum is thereby represented as, for example,

$$\hat{X} = \int d[q'] \, d[p'] \, U(q'p') \, X \, U(q'p')^{-1} \qquad (6.134)$$

where, in the notation of (6.57) ,

$$U(q'p') = U(q'p'0) = e^{i(pq'-p'q)} \quad . \qquad (6.135)$$

(The transformations described by the parameter λ are without effect here). The integrations are extended over the infinite spectral range of Hermitian canonical variables. As we can recognize directly from the group multiplication law (6.58), the operator $\hat{X}$ commutes with every unitary operator $U(q'p')$ Hence it commutes with both sets of canonical variables, q and p , and for a system described by variables of the first kind, $\hat{X}$ must be a multiple

---

$$\left(\tfrac{1}{2}y^{\dagger'} \frac{\partial}{\partial y^{\dagger'}}\right)^2 + \tfrac{1}{2}y^{\dagger'} \frac{\partial}{\partial y^{\dagger'}} \quad ,$$

and any function of the $y^{\dagger'}$ that is homogeneous of integral degree n provides an eigenfunction, which is associated with the eigenvalue $j(j+1)$, $j = n/2$. One recognizes that the differential operators refer to an equivalent system described by two complementary pairs of non-Hermitian variables of the first kind, and

$$\mathbf{J} = y^{\dagger}\tfrac{1}{2}\sigma y \quad .$$

This equivalence was used for a systematic development of the theory of angular momentum in a paper of the unit operator. This result can also be derived through an explicit construction of the matrix representing $\hat{X}$ in some canonical representation. We first observe that

$$\begin{aligned} \langle q''|U(q'p') &= \langle q''|e^{-ip'q}\, e^{ipq'}\, e^{-\frac{1}{2}ip'q'} \\ &= e^{-ip'(q''+\frac{1}{2}q')}\, \langle q'+q''| \quad , \end{aligned} \tag{6.136}$$

and therefore

$$\begin{aligned} \langle q''|\hat{X}|q'''\rangle &= \int d[q']\, d[p']\, e^{-ip'(q''-q''')} \langle q'+q''|X|q'+q'''\rangle \\ &= \delta[q''-q''']\ \mathrm{tr}\, X = \langle q''|1|q'''\rangle\ \mathrm{tr}\, X \ . \end{aligned} \tag{6.137}$$

We have thus shown that

$$\int d[q']\, d[p']\, U(q'p')\, X\, U(q'p')^{-1} = \mathrm{tr}\, X \quad , \tag{6.138}$$

where the right side is a multiple of the unit operator. If X is chosen as a measurement symbol, $M(a')$, this equation reads

$$1 = \int d[q']\, d[p']\, U(q'p')\, M(a')\, U(q'p')^{-1} \quad , \tag{6.139}$$

---

that was written in 1951, but remained unpublished. It is now available in the collection "Quantum Theory of Angular Momentum" edited by L.C. Biedenharn and H. Van Dam, Academic Press, 1965.

and multiplication on the right by $XM(a')$, followed by summation over a', converts our result into

$$X = \int d[q']\, d[p']\, U(q'p')\, \mathrm{tr}\left(U(q'p')^{-1} X\right) , \quad (6.140)$$

which is the explicit exhibition of any operator as a function of the fundamental dynamical variables of the first kind. If the operator X of (6.138) is replaced by such a function, $F(q, p)$, that formula becomes

$$1\, \mathrm{tr}\, F(q, p) = \int d[q']\, d[p']\, F(q+q', p+p') \quad . \quad (6.141)$$

As an application of the latter form, we have

$$\mathrm{tr}\, U(q'', p'') = \int d[q']\, d[p']\, e^{i[(p+p')q''-p''(q+q')]}$$
$$= \delta[q'']\, \delta[p''] \quad , \quad (6.142)$$

or, according to the multiplication property (6.58),

$$\mathrm{tr}\left(U(q'p')^{-1} U(q''p'')\right) = \delta[q'-q'']\, \delta[p'-p''] \quad . \quad (6.143)$$

The statements contained in (6.140) and (6.143) can be regarded as assertions of the completeness and orthonormality of the operator basis provided by the continuous set of unitary operator functions of the variables of the first kind, $U(q'p')$.

6.14 VARIABLES OF THE SECOND KIND

The general formal analogy between the two types of dynamical variables, together with the specific difference associated with the evaluation of traces, correctly suggests that the statements of Eqs. (6.138), (6.140-3) are applicable to variables of the second kind if the operation $\operatorname{tr} \ldots$ is replaced by $\operatorname{tr} \rho \ldots$. But a word of caution is needed. The eigenvalues that appear in $U(q'p')$ are anticommutative with the variables of the second kind and thus contain the factor ρ , whereas the eigenvalues involved in the integrations and in the delta functions are entirely elements of the external algebra. Incidentally, the general trace formula (4.179) emerges from (6.138) and its analogue for variables of the second kind on forming the $\langle p'=0| \; |q'=0\rangle$ matrix element. It is also worth noting that our various results could be freed from explicit reference to complementary variables. We shall make limited use of this possibility to transform the statement for variables of the second kind from their non-Hermitian canonical versions to forms appropriate for Hermitian canonical variables. This will be accomplished by using the transformation (3.116) for operators, together with the analoqous one for eigenvalues, which yields, for example,

$$U(q'p') = e^{i(pq'-p'q)} = e^{\xi'\xi} \tag{6.144}$$

or

$$U(\xi') = \prod_{\kappa=1}^{2n} \left(1 + \xi'_\kappa \xi_\kappa\right) \quad . \tag{6.145}$$

Similarly

$$\begin{aligned} \int d[q']\, d[p'] &= \int d[\xi'] \\ &= i^{-n} \prod_{\kappa=1}^{2n} \frac{\partial_\ell}{\partial \xi'_\kappa} = i^{n} \prod_{\kappa=1}^{2n}{}^{T} \frac{\partial_\ell}{\partial \xi'_\kappa} \end{aligned} \tag{6.146}$$

and

$$\begin{aligned} \delta[q']\, \delta[p'] &= \delta[\xi] \\ &= i^{-n} \prod_{\kappa=1}^{2n} \xi'_\kappa \quad . \end{aligned} \tag{6.147}$$

When using this notation one must not confuse the

symbol ξ_κ' $(\xi_\kappa'^2=0)$ with an eigenvalue of the Hermitian operator ξ_κ .

6.15 REFLECTION OPERATOR

If the integration process (6.109) is applied to the operators $U(q'p'\lambda)$, referring to variables of the first kind, we obtain zero as a result of the λ integration. Hence there is no state invariant under all the operations of the special canonical group. But, if the λ integration is omitted, we are led to consider the Hermitian operator

$$R = \frac{1}{2^n} \int d[q'] \, d[p'] \, U(q'p') \quad , \tag{6.148}$$

which is such that

$$\begin{aligned} R\, U(q''p'') &= \frac{1}{2^n} \int d[q'] \, d[p'] \, U(q'p') \, e^{\frac{1}{2}i(p'q''-p''q')} \\ &= U(-q'' , -p'') \, R \quad , \end{aligned} \tag{6.149}$$

and

$$\begin{aligned} R^2 &= \int \frac{d[q']}{2^n} \frac{d[p']}{2^n} \, U(q'p') \, \delta[\tfrac{1}{2}q'] \, \delta[\tfrac{1}{2}p'] \\ &= 1 \quad . \end{aligned} \tag{6.150}$$

Thus, R anticommutes with each variable of the first kind,

$$Rq\,R^{-1} = -q \quad , \quad Rp\,R^{-1} = -p \quad , \tag{6.151}$$

and correspondingly

$$R|q''\rangle = \frac{1}{2^n}\int d[q']\,d[p']\,|q''-q'\rangle\,e^{-ip'(q''-\frac{1}{2}q')}$$

$$= \int \frac{d[q']}{2^n}\,|q''-q'\rangle\,\delta[q''-\tfrac{1}{2}q'] = |-q''\rangle \quad . \tag{6.152}$$

The now familiar operator possessing these properties for the variables of the second kind is similarly produced by

$$\rho = 2^n \int d[\xi']\,U(\xi') \quad . \tag{6.153}$$

6.16 FINITE OPERATOR BASIS

The formal expressions of completeness and orthogonality for the operator basis of the variables of the second kind, which are comprised in the various aspects of the $U(\xi)$,

$$1 \text{ tr } X = \int d[\xi'] \, U(\xi') X \rho \, U(\xi')^{-1} \quad ,$$

$$X = \int d[\xi'] \, U(\xi'] \text{ tr } \left(\rho \, U(\xi')^{-1} X\right) \quad , \qquad (6.154)$$

$$\text{tr } \left(\rho \, U(\xi')^{-1} U(\xi'')\right) = \delta[\xi' - \xi''] \quad ,$$

can be freed of explicit reference to the special canonical group. We first recognize that $U(\xi')$ is the generating function for the 2^{2n} distinct elements of the operator algebra. These we define more precisely as

$$\alpha\{\nu_1 , \dots , \nu_{2n}\} = i^{-\frac{1}{2}\nu(\nu-1)} \prod_{\kappa=1}^{2n} \left(2^{\frac{1}{2}}\xi_\kappa\right)^{\nu_\kappa} \qquad (6.155)$$

where each ν_κ assumes the value 0 or 1, and

$$\nu = \sum_{\kappa=1}^{2n} \nu_\kappa \quad . \qquad (6.156)$$

With this definition the Hermitian operators $\alpha\{\nu\}$ possess unit squares and, in particular,

$$\alpha\{0\} = 1 \quad , \quad \alpha\{1\} = \rho \quad . \qquad (6.157)$$

If we use the notation $\alpha'\{\nu\}$ to designate the similar products formed from the eigenvalues ξ'_κ, we have

$$U(\xi') = \sum_{\{\nu\}} 2^{-\nu} \alpha'\{\nu\} \alpha\{\nu\} \quad . \tag{6.158}$$

The product of two operators, with indices $\{\nu\}$ and $\{1 - \nu\}$ equals the operator ρ, to within a phase factor,

$$\alpha\{\nu\} \alpha\{1-\nu\} = \varepsilon\{\nu\} \rho \quad ,$$
$$\varepsilon\{\nu\} \varepsilon\{1-\nu\} = (-1)^{\nu} \varepsilon\{\nu\}^2 = 1 \quad , \tag{6.159}$$

and therefore

$$\rho \, U(\xi')^{-1} = U(\xi')\rho = \sum_{\{\nu\}} 2^{-(2n-\nu)} \alpha'\{1-\nu\} \varepsilon\{\nu\} \alpha\{\nu\} \; . \tag{6.160}$$

The integration symbol stands for differentiation with respect to each ξ'_ν, and thus the terms that contribute to the integrals are of the form

$$\int d[\xi] \, \alpha'\{\nu\} \alpha'\{1-\nu\} = i^n \prod_{\kappa=1}^{2n}{}^{T} \frac{\partial_\ell}{\partial \xi'_\kappa} \varepsilon\{\nu\} \, i^{-n} 2^n \left(\prod \xi'_\kappa\right)$$
$$= 2^n \varepsilon\{\nu\} \quad . \tag{6.161}$$

But one must also recall that the anticommutative operator ρ is to be separated from the eigenvalues prior to integration. This has no explicit effect if the number of eigenvalues in the individual terms is even, corresponding to $\nu = \sum \nu_\kappa$ an even integer. For odd ν, however, the additional factor of ρ multiplying $\alpha\{\nu\}$ induces

$$\rho\, \alpha\{\nu\} = \varepsilon\{\nu\}\, \alpha\{1-\nu\} \quad . \tag{6.162}$$

The net result has the same form in either circumstance, leading to the following expression of completeness for the 4^n dimensional operator basis $\alpha\{\nu\}$

$$\begin{aligned} 1 \operatorname{tr} X &= \frac{1}{2^n} \sum_{\{\nu\}} \alpha\{\nu\}\, X\, \alpha\{\nu\} \\ X &= \frac{1}{2^n} \sum_{\{\nu\}} \alpha\{\nu\} \operatorname{tr}\left(\alpha\{\nu\}\, X\right) \ . \end{aligned} \tag{6.163}$$

The orthonormality property

$$\frac{1}{2^n} \operatorname{tr}\left(\alpha\{\nu\}\, \alpha\{\nu'\}\right) = \delta\{\nu\ ,\ \nu'\} \tag{6.164}$$

can be inferred from the completeness and linear independence of the $\alpha\{\nu\}$ or obtained directly from (6.154) on remarking that

$$\delta[\xi'-\xi''] = \frac{1}{2^n} \sum_{\{\nu\}} \varepsilon\{\nu\}\, \alpha'\{\nu\}\, \alpha''\{1-\nu\} \quad . \tag{6.165}$$

6.17 ADDENDUM: DERIVATION OF THE ACTION PRINCIPLE†

† Reproduced from the Proceedings of the National Academy of Sciences, Vol. 46, pp. 893 - 897 (1960).

We are now going to examine the construction of finite unitary transformations from infinitesimal ones for a physical system of n continuous degrees of freedom. Thus, all operators are functions of the n pairs of complementary variables q_k, p_k, which we denote collectively by x. Let us consider a continuous set of unitary operators labeled by a single parameter, $U(\tau)$. The change from τ to $\tau + d\tau$ is the infinitesimal transformation

$$1 + id\tau G(x, \tau),$$

which includes a possible explicit τ dependence of the generator, and

$$\begin{aligned} U(\tau + d\tau) &= [1 + id\tau G(x, \tau)]U(\tau) \\ &= U(\tau)[1 + id\tau G(x(\tau), \tau)] \end{aligned}$$

where

$$x(\tau) = U(\tau)^{-1}xU(\tau)$$

are the fundamental quantum variables of the system for the description produced by the transformation $U(\tau)$. The accompanying state transformations are indicated by

$$\langle a' | U(\tau) = \langle a'\tau |$$
$$U(\tau)^{-1} | b'\rangle = | b'\tau\rangle.$$

A useful representation of the unitary transformation is given by the transformation function

$$\langle a'\tau_1 | b'\tau_2\rangle = \langle a' | U(\tau_1)U(\tau_2)^{-1} | b'\rangle$$

which includes

$$\langle a'\tau | b'0\rangle = \langle a' | U(\tau) | b'\rangle,$$

the matrix of $U(\tau)$ in the arbitrary ab representation. The relation between infinitesimally neighboring values of τ is indicated by

$$\begin{aligned}\langle a'\tau + d\tau | b'\tau\rangle &= \langle a'\tau | [1 + id\tau G(x(\tau), \tau)] | b'\tau\rangle \\ &= \langle a' | [1 + id\tau G(x, \tau)] | b'\rangle.\end{aligned}$$

The general discussion of transformation functions indicates that the most compact characterization is a differential one. Accordingly, we replace this explicit statement of the transformation function $\langle a'\tau + d\tau | b'\tau\rangle$ by a differential description in which the guiding principle will be the maintenance of generality by avoiding considerations that refer to specific choices of the states a' and b'. We note first that the transformation function depends upon the parameters τ, $\tau + d\tau$ and upon the form of the generator $G(x, \tau)$. Infinitesimal changes in these aspects $[\delta'']$ induce the alteration

$$\begin{aligned}\delta''\langle \tau + d\tau | \tau\rangle &= i\langle | \delta''[d\tau G(x, \tau)] | \rangle \\ &= i\langle \tau + d\tau | \delta''[d\tau G(x(\tau), \tau)] | \tau\rangle,\end{aligned}$$

where the omission of the labels a', b' emphasizes the absence of explicit reference to these states. Yet some variation of the states must be introduced if a sufficiently complete characterization of the transformation function is to be obtained. For this purpose we use the infinitesimal transformations of the special canonical group, performed independently on the states associated with parameters τ and $\tau + d\tau[\delta']$. Thus

$$\delta'\langle \tau + d\tau | = i\langle \tau + d\tau | G_s(\tau + d\tau)$$
$$\delta' | \tau\rangle = -iG_s(\tau) | \tau\rangle$$

in which the infinitesimal generators are constructed from the operators appropriate to the description employed for the corresponding vectors, namely $x(\tau + d\tau)$ and $x(\tau)$. It is convenient to use the symmetrical generator $G_{q,\,p}$ which produces changes of the variables x by $^1/_2\delta x$. Then

$$G_s(\tau) = {}^1/_2 p(\tau)\delta q(\tau) - \delta p(\tau)q(\tau)$$

which, with the similar expression for $G_s(\tau + d\tau)$, gives

$$\delta'\langle \tau + d\tau | \tau\rangle = i\langle \tau + d\tau | [G_s(\tau + d\tau) - G_s(\tau)] | \tau\rangle,$$

where

$$G_s(\tau + d\tau) - G_s(\tau) = {}^1/_2[p(\tau + d\tau)\delta q(\tau + d\tau) + \delta p(\tau)q(\tau) - p(\tau)\delta q(\tau) - \delta p(\tau + d\tau)q(\tau + d\tau)],$$

and the $\delta x(\tau)$, $\delta x(\tau + d\tau)$ are independent arbitrary infinitesimal numbers upon which we impose the requirement of continuity in τ.

The infinitesimal unitary transformation that relates $x(\tau)$ and $x(\tau + d\tau)$ is obtained from

$$x(\tau + d\tau) = U(\tau + d\tau)^{-1}xU(\tau + d\tau)$$
$$= [1 - id\tau G(x(\tau), \tau)]\, x(\tau)[1 + id\tau G(x(\tau), \tau)]$$

as

$$x(\tau + d\tau) = x(\tau) - (1/i)[x(\tau), d\tau G(x(\tau), \tau)].$$

Accordingly, one can write

$$G_s(\tau + d\tau) - G_s(\tau) = {}^1/_2[p(\tau)\delta q(\tau + d\tau) + \delta p(\tau)q(\tau + d\tau) - p(\tau + d\tau)\delta q(\tau) - \delta p(\tau + d\tau)q(\tau)] - (1/i)[p(\tau)\delta q(\tau) - \delta p(\tau)q(\tau),$$
$$d\tau G(x(\tau),\tau)] = \delta'[{}^1/_2(p(\tau)q(\tau + d\tau) - p(\tau + d\tau)q(\tau))] + (1/i)[d\tau G(x(\tau), \tau), G_q(\tau) + G_p(\tau)]$$

or

$$G_s(\tau + d\tau) - G_s(\tau) = \delta'[{}^1/_2(p(\tau)q(\tau + d\tau) - p(\tau + d\tau)q(\tau)) + d\tau G(x(\tau), \tau)]$$

in which δ' is used here to describe the change of q, p by δq and δp, occurring independently but continuously at τ and $\tau + d\tau$. The two species of variation can now be united: $\delta = \delta' + \delta''$, and

$$\delta\langle \tau + d\tau | \tau \rangle = i\langle \tau + d\tau | \delta[W] | \tau \rangle,$$

where

$$W(\tau + d\tau, \tau) = {}^1/_2(p(\tau)q(\tau + d\tau) - p(\tau + d\tau)q(\tau)) + d\tau G(x(\tau), \tau)$$
$$= {}^1/_2(pdq - dpq) + d\tau G.$$

Our result is a specialization of the general differential characterization of transformation functions whereby, for a class of alterations, the infinitesimal operator δW is derived as the variation of a single operator W. This is a quantum action principle[3] and W is the action operator associated with the transformation.

We can now proceed directly to the action principle that describes a finite unitary transformation,

$$\delta\langle \tau_1 | \tau_2 \rangle = i\langle \tau_1 | \delta[W_{12}] | \tau_2 \rangle$$

for multiplicative composition of the individual infinitesimal transformation functions is expressed by addition of the corresponding action operators

$$W_{12} = \sum_{\tau_2}^{\tau_1} W(\tau + d\tau, \tau)$$
$$= \int_{\tau_2}^{\tau_1} [{}^1/_2(p(\tau)dq(\tau) - dp(\tau)q(\tau)) + d\tau G(x(\tau), \tau)].$$

As written, this action operator depends upon all operators $x(\tau)$ in the τ interval between τ_1 and τ_2. But the transformations of the special canonical group, applied to $\langle \tau_1 | \tau_2 \rangle$, give

$$\delta' \langle \tau_1 | \tau_2 \rangle = i \langle \tau_1 | [G_s(\tau_1) - G_s(\tau_2)] | \tau_2 \rangle$$

which is to say that $\delta' W_{12}$ does not contain operators referring to values of τ in the open interval between τ_1 and τ_2, or that W_{12} is stationary with respect to the special variations of $x(\tau)$ in that interval. Indeed, this principle of stationary action, the condition that a finite unitary transformation emerge from the infinitesimal ones, asserts of $q(\tau)$, $p(\tau)$ that

$$\frac{dq}{d\tau} = -\frac{\partial G}{\partial p}, \qquad \frac{dp}{dt} = \frac{\partial G}{\partial q},$$

which are immediate implications of the various infinitesimal generators.

The use of a single parameter in this discussion is not restrictive. We have only to write

$$G(x, \tau) = \sum_{k=1}^{p} G_k(x, \tau) \frac{d\tau_k(\tau)}{d\tau}$$

with each $d\tau_k/d\tau$ given as an arbitrary function of τ, and then regard the transformation as one with p parameters, conducted along a particular path in the parameter space that is specified by the p functions of a path parameter, $\tau_k(\tau)$. Now

$$W_{12} = \int_{\tau_2}^{\tau_1} [{}^1/_2(pdq - dpq) + \sum_k G_k(x(\tau), \tau) d\tau_k]$$

is the action operator for a transformation referring to a prescribed path and generally depends upon that path. If we consider an infinitesimal path variation with fixed end points we find that

$$\delta_{\text{path}} W_{12} = \int_{\tau_2}^{\tau_1} \sum_{kl} {}^1/_2(\delta\tau_k d\tau_l - \delta\tau_l d\tau_k) R_{kl}$$

where

$$R_{kl} = \frac{\partial}{\partial \tau_k} G_l - \frac{\partial}{\partial \tau_l} G_k + \frac{1}{i} [G_k, G_l]$$
$$= -R_{lk}.$$

The vanishing of each of these operators is demanded if the transformation is to be independent of path. When the operators $G_k(x, \tau)$ can be expressed as a linear combination of an equal number of operators that are not explicit functions of the parameters, $G_a(x)$, the requirement of path independence yields the previously considered conditions for the formation of a group.

We now have the foundations for a general theory of quantum dynamics and canonical transformations, at least for systems with continuous degrees of freedom. The question is thus posed whether other types of quantum variables can also be employed in a quantum action principle.

* Publication assisted by the Office of Scientific Research, United States Air Force, under contract number AF49(638)-589.

[1] These PROCEEDINGS, **45,** 1542 (1959); **46,** 257 (1960); and **46,** 570 (1960).

[2] The group can be obtained directly as the limit of the finite order group associated with each ν. That group, of order ν^3, is generated by U, V, and the νth root of unity given by $VUV^{-1}U^{-1}$. Some aspects of the latter group are worthy of note. There are $\nu^2 + \nu - 1$ classes, the commutator subgroup is of order ν, and the order of the corresponding quotient group, ν^2, is the number of inequivalent one-dimensional representations. The remaining $\nu - 1$ matrix representations must be of dimensionality ν, $\nu^3 = \nu^2 + (\nu - 1)\nu^2$, and differ only in the choice of the generating νth root of unity. That choice is already made in the statement of operator properties for U and V and there can be only one irreducible matrix representation of these operators, to within the freedom of unitary transformation.

[3] In earlier work of the author, for example *Phys. Rev.*, **91,** 713 (1953), the quantum action principle has been postulated rather than derived.

6.18 ADDENDUM CONCERNING THE SPECIAL CANONICAL GROUP†

---

† Reproduced from the Proceedings of the National Academy of Sciences, Vol. 46, pp. 1401-1415 (1960).

Reprinted from the Proceedings of the NATIONAL ACADEMY OF SCIENCES
Vol. 46, No. 10, pp. 1401–1415. October, 1960.

*THE SPECIAL CANONICAL GROUP**

BY JULIAN SCHWINGER

HARVARD UNIVERSITY

Communicated August 30, 1960

This note is concerned with the further development and application of an operator group described in a previous paper.[1] It is associated with the quantum degree of freedom labeled $\nu = \infty$ which is characterized by the complementary pair of operators q, p with continuous spectra.[2] The properties of such a degree of freedom are obtained as the limit of one with a finite number of states, specifically given by a prime integer ν. We recall that unitary operators U and V obeying

$$U^\nu = V^\nu = 1, \; VU = e^{2\pi i/\nu}UV$$

define two orthonormal coordinate systems $\langle u^k|$ and $\langle v^k|$, where

$$u^k = v^k = e^{2\pi ik/\nu}$$

and

$$\langle u^k|v^l\rangle = \nu^{-1/2}e^{2\pi ikl/\nu}.$$

For any prime $\nu > 2$ we can choose the integers k and l to range from $-{}^1/_2(\nu - 1)$ to ${}^1/_2(\nu - 1)$, rather than from 0 to $\nu - 1$. An arbitrary state Ψ can be represented alternatively by the wave functions

$$\psi(u^k) = \langle u^k|\Psi, \; \psi(v^k) = \langle v^k|\Psi$$

where

$$\Psi^\dagger\Psi = \sum_k |\psi(u^k)|^2 = \sum_k |\psi(v^k)|^2,$$

and the two wave functions are reciprocally related by

$$\psi(u^k) = \sum_l \nu^{-1/2} e^{2\pi ikl/\nu} \psi(v^l)$$

$$\psi(v^l) = \sum_k \nu^{-1/2} e^{-2\pi ikl/\nu} \psi(u^k).$$

We now shift our attention to the Hermitian operators q, p defined by

$$U = e^{i\epsilon q}, \quad V = e^{i\epsilon p}, \quad \epsilon = (2\pi/\nu)^{1/2}$$

and the spectra

$$q', p' = \epsilon[-1/2(\nu - 1) \;..\; 0 \;..\; 1/2(\nu - 1)].$$

Furthermore, we redefine the wave functions so that

$$\psi(u^k) = \epsilon^{1/2}\psi(q'), \; \psi(v^k) = \epsilon^{1/2}\psi(p')$$

where $\epsilon = \Delta q' = \Delta p'$, the interval between adjacent eigenvalues. Then we have

$$\Psi^\dagger\Psi = \sum_{q'} \Delta q'|\psi(q')|^2 = \sum_{p'} \Delta p'|\psi(p')|^2$$

and

$$\psi(q') = \sum_{p'} \Delta p'(2\pi)^{-1/2}\, e^{iq'p'}\, \psi(p')$$

$$\psi(p') = \sum_{q'} \Delta q'(2\pi)^{-1/2}\, e^{-iq'p'}\, \psi(q').$$

As ν increases without limit the spectra of q and p become arbitrarily dense and the eigenvalues of largest magnitude increase indefinitely. Accordingly we must restrict all further considerations to that physical class of states, or physical subspace of vectors, for which the wave functions $\psi(q')$ and $\psi(p')$ are sufficiently well behaved with regard to continuity and the approach of the variable to infinity that a uniform transition to the limit $\nu = \infty$ can be performed, with the result

$$\Psi^\dagger\Psi = \int_{-\infty}^{\infty} dq' |\psi(q')|^2 = \int_{-\infty}^{\infty} dp' |\psi(p')|^2$$

and

$$\psi(q') = \int_{-\infty}^{\infty} dp'\ (2\pi)^{-1/2}\, e^{iq'p'}\, \psi(p')$$

$$\psi(p') = \int_{-\infty}^{\infty} dq'\ (2\pi)^{-1/2}\, e^{-iq'p'}\, \psi(q').$$

We shall not attempt here to delimit more precisely the physical class of states. Note however that the reciprocal relation between wave functions can be combined into

$$\psi(q') = \int_{-\infty}^{\infty} \frac{dp'}{2\pi}\, e^{iq'p'} \int_{-\infty}^{\infty} dq''\, e^{-iq''p'}\, \psi(q'')$$

which must be an identity for wave functions of the physical class when the operations are performed as indicated. There will also be a class of functions $K(p', \epsilon)$, such that

$$K(p', \epsilon) \rightarrow \begin{cases} 1, & \epsilon \rightarrow 0 \\ 0, & |p'| \rightarrow \infty \end{cases}$$

and

$$\psi(q') = \lim_{\epsilon \rightarrow 0} \int_{-\infty}^{\infty} \frac{dp'}{2\pi}\, e^{iq'p'} K(p', \epsilon) \int_{-\infty}^{\infty} dq''\, e^{-iq''p'}\, \psi(q'')$$

$$= \lim_{\epsilon \rightarrow 0} \int_{-\infty}^{\infty} dq'' \left[\int_{-\infty}^{\infty} \frac{dp'}{2\pi}\, K(p', \epsilon)\, e^{ip'(q'-q'')}\right] \psi(q'').$$

This is what is implied by the symbolic notation (Dirac)

$$\psi(q') = \int_{-\infty}^{\infty} dq''\delta(q' - q'')\psi(q'')$$

$$\delta(q' - q'') = \int_{-\infty}^{\infty} \frac{dp'}{2\pi}\, e^{ip'(q'-q'')}$$

We shall also use the notation that is modeled on the discrete situation, with integrals replacing summations, as in

$$\langle q'|q''\rangle = \int_{-\infty}^{\infty} \langle q'|p'\rangle dp' \langle p'|q''\rangle = \delta(q' - q''),$$

with

$$\langle q'|p'\rangle = \langle p'|q'\rangle^* = (2\pi)^{-1/2}\, e^{iq'p'}.$$

There are other applications of the limit $\nu \to \infty$. The reciprocal property of the operators U and V is expressed by

$$\langle u^k| V = \langle u^{k+1}|, \ \langle v^k| U^{-1} = \langle v^{k+1}|$$

or

$$\langle q'| e^{i\epsilon p} = \langle q' + \epsilon|, \ \langle p'| e^{-i\epsilon q} = \langle p' + \epsilon|,$$

with an exception when q' or p' is the greatest eigenvalue, for then $q' + \epsilon$ or $p' + \epsilon$ is identified with the least eigenvalue, $-q'$ or $-p'$. We write these relations as wave function statements, of the form

$$\frac{1}{\epsilon} [\langle q' + \epsilon| \Psi - \langle q'| \Psi] = \langle q'| \frac{1}{\epsilon} (e^{i\epsilon p} - 1)\Psi$$

and

$$\frac{1}{\epsilon} [\langle p' + \epsilon| \Psi - \langle p'| \Psi] = \langle p'| \frac{1}{\epsilon} (e^{-i\epsilon q} - 1)\Psi.$$

In the transition to the limit $\nu = \infty$, the subspace of physical vectors Ψ is distinguished by such properties of continuity and behavior at infinity of the wave functions that the left-hand limits $\epsilon \to 0$, exist as derivatives of the corresponding wave function. We conclude, for the physical class of states, that

$$\frac{1}{i} \frac{\partial}{\partial q'} \langle q'| \Psi = \langle q'| p\Psi$$

and

$$i \frac{\partial}{\partial p'} \langle p'| \Psi = \langle p'| q\Psi.$$

It will also be evident from this application of the limiting process to the unitary operators $\exp(i\epsilon q)$, $\exp(i\epsilon p)$ that a restriction to a physical subspace is needed for the validity of the commutation relation

$$[q, p] = i.$$

The elements of an orthonormal operator basis are given by

$$\nu^{-1/2} e^{\pi imn/\nu} U^m V^n = \nu^{-1/2} e^{-(\pi imn/\nu)} V^n U^m$$

or $\nu^{-1/2} U(q'p')$, with

$$U(q'p') = e^{1/2ip'q'} e^{ipq'} e^{-ip'q} = e^{-1/2ip'q'} e^{-ip'q} e^{ipq'}$$
$$= (\nu \to \infty)\ e^{i(pq' - p'q)}.$$

Thus, since $\nu^{-1} = \Delta q' \Delta p'/2\pi$, we have, for an arbitrary function $f(q'p')$ of the discrete variables q', p',

$$\sum_{q''p''} \frac{\Delta q'' \Delta p''}{2\pi} tr\ U(q'p')^{\dagger} U(q''p'') f(q''p'') = f(q'p').$$

In the limit $\nu = \infty$ there is a class of functions $f(q'p')$ such that

$$tr\ U(q'p')^\dagger \int_{-\infty}^{\infty} \frac{dq''dp''}{2\pi} U(q''p'')f(q''p'') = f(q'p'),$$

which we express symbolically by

$$tr\ U(q'p')^\dagger U(q''p'') = 2\pi\delta(q' - q'')\delta(p' - p'').$$

In particular,

$$\begin{aligned} tr\ e^{i(pq'-p'q)} &= 2\pi\delta(q')\delta(p') \\ &= \int_{-\infty}^{\infty} \frac{dqdp}{2\pi} e^{i(pq'-p'q)} \end{aligned}$$

where q and p on the right-hand side are numerical integration variables.

The completeness of the operator basis $U(q'p')$ is expressed by

$$\int_{-\infty}^{\infty} \frac{dq'dp'}{2\pi} U(q'p')\ X\ U(q'p')^\dagger = 1\ tr\ X.$$

The properties of the $U(q'p')$ basis are also described, with respect to an arbitrary discrete operator basis $X(\alpha)$, by

$$\int_{-\infty}^{\infty} \frac{dq'dp'}{2\pi} \langle\alpha|\, q'p'\rangle\, \langle q'p'|\, \alpha'\rangle = \delta(\alpha, \alpha')$$

and

$$\sum_\alpha \langle q'p'|\, \alpha\rangle\, \langle\alpha|\, q''p''\rangle = 2\pi\delta(q' - q'')\delta(p' - p'').$$

When X is given as $F(q, p)$, the operations of the special canonical group can be utilized to bring the completeness expression into the form

$$\int_{-\infty}^{\infty} \frac{dq'dp'}{2\pi} F(q + q', p + p') = 1\ tr\ F(q, p).$$

This operator relation implies a numerical one if it is possible to order $F(q, p)$, so that all q operators stand to the left, for example, of the operator p: $F(q; p)$. Then the evaluation of the $\langle q' = 0| \quad |p' = 0\rangle$ matrix element gives

$$tr\ F(q, p) = \int_{-\infty}^{\infty} \frac{dqdp}{2\pi} F(q; p)$$

which result also applies to a system with n continuous degrees of freedom if it is understood that

$$\frac{dqdp}{2\pi} = \prod_{k=1}^{n} \frac{dq_k dp_k}{2\pi}.$$

As an example of this ordering process other than the one already given by $tr\ U(q'p')$, we remark that

$$e^{-1/2(q^2+p^2)\beta} = (\cosh\beta)^{-\frac{1}{2}}\, e^{-\frac{1}{2}q^2 \tanh\beta}\, e^{iq;\ p(\operatorname{sech}\beta - 1)}\, e^{-1/2p^2 \tanh\beta}$$

where

$$e^{a;\ b} = \sum_0^\infty \frac{1}{n!} a^n b^n,$$

and therefore

$$tr\, e^{-1/2(q^2+p^2)\beta} = \int_{-\infty}^{\infty} \frac{dqdp}{2\pi} (\cosh \beta)^{-1/2}\, e^{-1/2(q^2+p^2)\tanh \beta}\, e^{iqp\,(\operatorname{sech} \beta - 1)}$$

$$= \frac{1}{2 \sinh {}^1/_2\beta} = \sum_{n=0}^{\infty} e^{-(n+1/2)\beta},$$

which reproduces the well-known non-degenerate spectrum of the operator ${}^1/_2(q^2 + p^2)$.

Now we shall consider the construction of finite special canonical transformations from a succession of infinitesimal ones, as represented by the variation of a parameter τ. Let the generator of the transformation associated with $\tau \rightarrow \tau + d\tau$ be

$$d\tau G_s = d\tau(qP - pQ)$$

where $Q(\tau)$ and $P(\tau)$ are arbitrary numerical functions of τ. That is, the infinitesimal transformation is

$$q(\tau + d\tau) - q(\tau) = -\frac{1}{i}[q, d\tau G_s] = d\tau Q(\tau)$$

$$p(\tau + d\tau) - p(\tau) = -\frac{1}{i}[p, d\tau G_s] = d\tau P(\tau),$$

which implies the finite transformation

$$q(\tau_1) - q(\tau_2) = \int_{\tau_2}^{\tau_1} d\tau Q(\tau)$$

$$p(\tau_1) - p(\tau_2) = \int_{\tau_2}^{\tau_1} d\tau P(\tau).$$

Some associated transformation functions are easily constructed. We have

$$\langle \tau + d\tau| = \langle \tau|\,[1 + id\tau G(x(\tau), \tau)]$$

or

$$\frac{1}{i}\frac{\partial}{\partial \tau}\langle \tau| = \langle \tau|\,[q(\tau)P(\tau) - p(\tau)Q(\tau)],$$

and therefore

$$\frac{1}{i}\frac{\partial}{\partial \tau}\langle q'\tau|p'\tau_2\rangle$$

$$= \langle q'\tau|\,[q'P(\tau) - (p' + \int_{\tau_2}^{\tau} d\tau' P(\tau'))Q(\tau)]\,|p'\tau_2\rangle$$

which, in conjunction with the initial condition

$$\tau = \tau_2\colon\ \langle q'\tau|p'\tau_2\rangle = \langle q'|p'\rangle = (2\pi)^{-1/2} e^{iq'p'}$$

gives

$$\langle q'\tau_1|p'\tau_2\rangle^{QP} = (2\pi)^{-1/2} \exp\,[i(q'p' + q'\int_{\tau_2}^{\tau_1} d\tau P(\tau) - p'\int_{\tau_2}^{\tau_1} d\tau Q(\tau) - \int_{\tau_2}^{\tau_1} d\tau d\tau' Q(\tau)\eta_+(\tau - \tau')P(\tau'))]$$

with

$$\eta_+(\tau - \tau') = \begin{cases} 1, & \tau > \tau' \\ 0, & \tau < \tau' \end{cases}.$$

From this result we derive

$$\langle q'\tau_1 | q''\tau_2\rangle^{QP} = \int_{-\infty}^{\infty} \langle q'\tau_1 | p'\tau_2\rangle^{QP} dp' \langle p' | q''\rangle = \delta(q' - q'' - \int_{\tau_2}^{\tau_1} d\tau Q(\tau))\, e^{iq'\int d\tau P}\, e^{-i\int d\tau d\tau' Q\eta_+ P},$$

and

$$\langle p'\tau_1 | p''\tau_2\rangle^{QP} = \int_{-\infty}^{\infty} \langle p' | q'\rangle dq' \langle q'\tau_1 | p''\tau_2\rangle^{QP} = \delta(p' - p'' - \int_{\tau_2}^{\tau_1} d\tau P(\tau))\, e^{-ip''\int d\tau Q}\, e^{-i\int d\tau d\tau' Q\eta_+ P}$$

or, alternatively,

$$\delta(p' - p'' - \int d\tau P)\, e^{-ip'\int d\tau Q}\, e^{i\int d\tau d\tau' P\eta_+ Q},$$

on using the fact that

$$\eta_-(\tau - \tau') = 1 - \eta_+(\tau - \tau') = \eta_+(\tau' - \tau).$$

These transformation functions can also be viewed as matrix elements of the unitary operator, an element of the special canonical group, that produces the complete transformation. That operator, incidentally, is

$$\exp[i(q \int_{\tau_2}^{\tau_1} d\tau P(\tau) - p \int_{\tau_2}^{\tau_1} d\tau Q(\tau) - {}^1/_2 \int_{\tau_2}^{\tau_1} d\tau d\tau' Q(\tau)\epsilon(\tau - \tau')P(\tau'))]$$

where $\epsilon(\tau - \tau')$ is the odd step function

$$\epsilon = \eta_+ - \eta_-.$$

We can compute the trace of a transformation function, regarded as a matrix, and this will equal the trace of the associated unitary operator provided the otherwise arbitrary representation is not an explicit function of τ. Thus

$$tr\langle\tau_1 | \tau_2\rangle^{QP} = \int_{-\infty}^{\infty} dq' \langle q'\tau_1 | q'\tau_2\rangle^{QP} = \int_{-\infty}^{\infty} dp' \langle p'\tau_1 | p'\tau_2\rangle^{QP} = 2\pi\delta\,(\int_{\tau_2}^{\tau_1} d\tau Q(\tau))\delta(\int_{\tau_2}^{\tau_1} d\tau P(\tau))\, e^{-i\int d\tau d\tau' Q\eta_+ P}$$

where, in view of the delta function factors, $\eta_+(\tau - \tau')$ can be replaced by other equivalent functions, such as $\eta_+ - {}^1/_2 = {}^1/_2\epsilon$, or ${}^1/_2\epsilon(\tau - \tau') - (\tau - \tau')/T$, with $T = \tau_1 - \tau_2$. The latter choice has the property of giving a zero value to the double integral whenever $Q(\tau)$ or $P(\tau)$ is a constant. As an operator statement, the trace formula is the known result

$$tr\ e^{i(pq' - p'q)} = 2\pi\delta(q')\delta(p').$$

It is important to recognize that the trace, which is much more symmetrical than any individual transformation function, also implies specific transformation functions. Thus, let us make the substitutions

$$\left.\begin{array}{l} Q(\tau) \rightarrow Q(\tau) - (q' - q'')\delta(\tau - \tau_1 + \epsilon T) \\ P(\tau) \rightarrow P(\tau) + p'\delta(\tau - \tau_2 - \epsilon T) \end{array}\right\} ,\ \epsilon \rightarrow +0$$

and then indicate the effect of the additional localized transformations by the equivalent unitary operators, which gives

$$tr\langle\tau_1 |\ e^{i(q' - q'')p(\tau_1)}\, e^{ip'q(\tau_2)} | \tau_2\rangle^{QP} = \int_{-\infty}^{\infty} dq \langle q + q' - q''\quad \tau_1 | q\tau_2\rangle^{QP}\, e^{ip'q}.$$

Accordingly, if we also multiply by $\exp(-ip'q'')$ and integrate with respect to $dp'/2\pi$, what emerges is $\langle q'\tau_1 | q''\tau_2\rangle^{QP}$, as we can verify directly.

We shall find it useful to give an altogether different derivation of the trace formula. First note that

$$tr\langle\tau_1|(q(\tau_1) - q(\tau_2))|\tau_2\rangle = \int_{-\infty}^{\infty} dq'\langle q'\tau_1|(q' - q')|q'\tau_2\rangle = 0$$

and similarly

$$tr\langle\tau_1|(p(\tau_1) - p(\tau_2))|\tau_2\rangle = 0$$

which is a property of periodicity over the interval $T = \tau_1 - \tau_2$. Let us, therefore, represent the operators $q(\tau)$, $p(\tau)$ by the Fourier series

$$q(\tau) = q_0 + \sum_1^\infty (\pi n)^{-1/2}\left[q_n \cos\frac{2\pi n}{T}(\tau - \tau_2) + q_{-n}\sin\frac{2\pi n}{T}(\tau - \tau_2)\right]$$

$$p(\tau) = p_0 + \sum_1^\infty (\pi n)^{-1/2}\left[-p_n \sin\frac{2\pi n}{T}(\tau - \tau_2) + p_{-n}\cos\frac{2\pi n}{T}(\tau - \tau_2)\right],$$

where the coefficients are so chosen that the action operator for an arbitrary special canonical transformation

$$W_{12} = \int_{\tau_2}^{\tau_1} [{}^1/_2(pdq - dpq) + d\tau(qP - pQ)]$$

acquires the form

$$W_{12} = \sum_{-\infty}^{\infty}{}' p_n q_n + \sum_{-\infty}^{\infty} (q_n P_n - p_n Q_n).$$

Here the dash indicates that the term $n = 0$ is omitted, and

$$Q_0 = \int_{\tau_2}^{\tau_1} d\tau Q(\tau),\ P_0 = \int_{\tau_2}^{\tau_1} d\tau P(\tau)$$

$$Q_{n,\,-n} = (\pi n)^{-1/2}\int_{\tau_2}^{\tau_1} d\tau Q(\tau)(-\sin,\,\cos)\left(\frac{2\pi n}{T}(\tau - \tau_2)\right)$$

$$P_{n,\,-n} = (\pi n)^{-1/2}\int_{\tau_2}^{\tau_1} d\tau P(\tau)(\cos,\,\sin)\left(\frac{2\pi n}{T}(\tau - \tau_2)\right).$$

The action principle for the trace is

$$\delta\ tr\ \langle\tau_1|\tau_2\rangle = i\ tr\ \langle\tau_1|\delta[W_{12}]|\tau_2\rangle$$

and the principle of stationary action asserts that

$$Q_0\ tr\ \langle\tau_1|\tau_2\rangle^{QP} = P_0\ tr\ \langle\tau_1|\tau_2\rangle^{QP} = 0,$$

together with

$$tr\ \langle\tau_1|(q_n - Q_n)|\tau_2\rangle^{QP} = tr\ \langle\tau_1|(p_n + P_n)|\tau_2\rangle^{QP} = 0.$$

The first of these results implies that the trace contains the factors $\delta(Q_0)$ and $\delta(P_0)$. The dependence upon Q_n, P_n, $n \neq 0$, is then given by the action principle as

$$\frac{\partial}{\partial Q_n}(tr) = i\ tr\ \langle\tau_1|(-p_n)|\tau_2\rangle = i\ P_n\ (tr)$$

$$\frac{\partial}{\partial P_n}(tr) = i\ tr\ \langle\tau_1|q_n|\tau_2\rangle = iQ_n(tr).$$

Therefore

$$tr\ \langle\tau_1|\ \tau_2\rangle^{QP} = 2\pi\delta(Q_0)\delta(P_0)\ e^{i\sum_{-\infty}^{\infty}{}' Q_nP_n}$$

where the factor of 2π is supplied by reference to the elementary situation with constant $Q(\tau)$ and $P(\tau)$. We note, for comparison with previous results, that

$$Q_nP_n + Q_{-n}P_{-n} = -\frac{1}{\pi n}\int_{\tau_2}^{\tau_1} d\tau d\tau' Q(\tau)\sin\left(\frac{2\pi n}{T}(\tau - \tau')\right)P(\tau')$$

and

$$\sum_{1}^{\infty}\frac{1}{\pi n}\sin\frac{2\pi n}{T}(\tau - \tau') = {}^1/_2\,\epsilon(\tau - \tau') - \frac{\tau - \tau'}{T}.$$

The new formula for the trace can be given a uniform integral expression by using the representation

$$e^{iQP} = \int_{-\infty}^{\infty}\frac{dqdp}{2\pi}\ e^{i(pq+qP-pQ)}$$

for now we can write

$$tr\ \langle\tau_1|\ \tau_2\rangle^{QP} = \int d[q, p]\ e^{iW[q,\ p]},$$

where

$$d[q, p] = \prod_{-\infty}^{\infty}\frac{dq_ndp_n}{2\pi}$$

and $W[q, p]$ is the numerical function formed in the same way as the action operator,

$$W[q, p] = \sum_{-\infty}^{\infty}{}'\ p_nq_n + \sum_{-\infty}^{\infty}(q_nP_n - p_nQ_n).$$

Alternatively, we can use the Fourier series to define the numerical functions $q(\tau)$, $p(\tau)$. Then

$$W[q, p] = \int_{\tau_2}^{\tau_1}[{}^1/_2(pdq - dpq) + d\tau(qP - pQ)]$$

and $d[q, p]$ appears as a measure in the quantum phase space of the functions $q(\tau)$, $p(\tau)$.

It is the great advantage of the special canonical group that these considerations can be fully utilized in discussing arbitrary additional unitary transformations, as described by the action operator

$$W_{12} = \int_{\tau_2}^{\tau_1}[{}^1/_2(pdq - dpq) + d\tau(qP - pQ) + d\tau G(x(\tau), \tau)].$$

First, let us observe how an associated transformation function $\langle\tau_1|\ \tau_2\rangle_G{}^{QP}$ depends upon the arbitrary functions $Q(\tau)$, $P(\tau)$. The action principle asserts that

$$\delta QP\ \langle\tau_1|\ \tau_2\rangle = i\langle\tau_1|\int_{\tau_2}^{\tau_1} d\tau(q\delta P - p\delta Q)|\ \tau_2\rangle = i\int_{\tau_2}^{\tau_1} d\tau[\delta P(\tau)\langle\tau_1|\ q(\tau)|\ \tau_2\rangle - \delta Q(\tau)\langle\tau_1|\ p(\tau)|\ \tau_2\rangle]$$

which we express by the notation

$$-i\frac{\delta}{\delta P(\tau)}\langle\tau_1|\tau_2\rangle = \langle\tau_1|q(\tau)|\tau_2\rangle$$

$$i\frac{\delta}{\delta Q(\tau)}\langle\tau_1|\tau_2\rangle = \langle\tau_1|p(\tau)|\tau_2\rangle.$$

More generally, if $F(\tau')$ is an operator function of $x(\tau')$ but not of Q, P we have

$$\langle\tau_1|F(\tau')|\tau_2\rangle_G{}^{QP} = \langle\tau_1|\tau'\rangle_G{}^{QP} \times \langle\tau'|F(\tau')|\tau'\rangle \times \langle\tau'|\tau_2\rangle_G{}^{QP},$$

where $|\tau'\rangle \times \langle\tau'|$ symbolizes the summation over a complete set of states, and therefore

$$-i\frac{\delta}{\delta P(\tau)}\langle\tau_1|F(\tau')|\tau_2\rangle = \langle\tau_1|(q(\tau)F(\tau'))_{\text{o}}|\tau_2\rangle$$

$$i\frac{\delta}{\delta Q(\tau)}\langle\tau_1|F(\tau')|\tau_2\rangle = \langle\tau_1|(p(\tau)F(\tau'))_{\text{o}}|\tau_2\rangle.$$

Here $(\quad)_{\text{o}}$ is an ordered product that corresponds to the sense of progression from τ_2 to τ_1. If τ follows τ' in sequence, the operator function of τ stands to the left, while if τ precedes τ' the associated operator appears on the right. This description covers the two algebraic situations: $\tau_1 > \tau_2$, where we call the ordering positive, $(\quad)_+$, and $\tau_1 < \tau_2$, which produces negative ordering, $(\quad)_-$. For the moment take $\tau_1 > \tau_2$ and compare

$$i\frac{\delta}{\delta Q(\tau)}\langle\tau_1|q(\tau+0)|\tau_2\rangle = \langle\tau_1|q(\tau)p(\tau)|\tau_2\rangle$$

with

$$i\frac{\delta}{\delta Q(\tau)}\langle\tau_1|q(\tau-0)|\tau_2\rangle = \langle\tau_1|p(\tau)q(\tau)|\tau_2\rangle.$$

The difference of these expressions,

$$i\frac{\delta}{\delta Q(\tau)}\langle\tau_1|[q(\tau+0) - q(\tau-0)]|\tau_2\rangle = \langle\tau_1|[q(\tau), p(\tau)]|\tau_2\rangle,$$

refers on one side to the noncommutativity of the complementary variables q and p, and on the other to the "equation of motion" of the operator $q(\tau)$. According to the action principle

$$\frac{dq}{d\tau} + \frac{\partial G}{\partial p} = Q$$

and therefore

$$q(\tau+0) - q(\tau-0) = \underset{\epsilon\to 0}{\text{Lim}}\int_{\tau-\epsilon}^{\tau+\epsilon} d\tau'\left[-\frac{\partial G}{\partial p} + Q(\tau')\right]$$

which yields the expected result,

$$i\frac{\delta}{\delta Q(\tau)}\langle\tau_1|[q(\tau+0) - q(\tau-0)]|\tau_2\rangle = i\langle\tau_1|\tau_2\rangle.$$

Thus, through the application of the special canonical group, we obtain functional differential operator representations for all the dynamical variables. The general statement is

$$\langle\tau_1|\,F(q,\,p)_o|\,\tau_2\rangle = F\left(-i\frac{\delta}{\delta P},\, i\frac{\delta}{\delta Q}\right)\langle\tau_1|\,\tau_2\rangle$$

where $F(q, p)_o$ is an ordered function of the $q(\tau)$, $p(\tau)$ throughout the interval between τ_2 and τ_1, and, as the simple example of $q(\tau)p(\tau)$ and $p(\tau)q(\tau)$ indicates, the particular order of multiplication for operators with a common value of τ must be reproduced by a suitable limiting process from different τ values.

The connection with the previous considerations emerges on supplying G with a variable factor λ. For states at τ_1 and τ_2 that do not depend explicitly upon λ we use the action principle to evaluate

$$\frac{\partial}{\partial\lambda}\langle\tau_1|\,\tau_2\rangle_{\lambda G}{}^{QP} = i\langle\tau_1|\int_{\tau_2}^{\tau_1} d\tau G(qp\tau)\,|\,\tau_2\rangle_{\lambda G}{}^{QP} = i\int_{\tau_2}^{\tau_1} d\tau G\left(-i\frac{\delta}{\delta P},\, i\frac{\delta}{\delta Q},\,\tau\right)\langle\tau_1|\,\tau_2\rangle_{\lambda G}{}^{QP}.$$

The formal expression that gives the result of integrating this differential equation from $\lambda = 0$ to $\lambda = 1$ is

$$\langle\tau_1|\,\tau_2\rangle_G{}^{QP} = \exp\left[i\int_{\tau_2}^{\tau_1} d\tau G\left(-i\frac{\delta}{\delta P},\, i\frac{\delta}{\delta Q},\,\tau\right)\right]\langle\tau_1|\,\tau_2\rangle^{QP}$$

where the latter transformation function is that for $\lambda = 0$ and therefore refers only to the special canonical group. An intermediate formula, corresponding to $G = G_1 + G_2$, contains the functional differential operator constructed from G_1 acting on the transformation function associated with G_2. The same structure applies to the traces of the transformation functions. If we use the integral representation for the trace of the special canonical transformation function, and perform the differentiations under the integration sign, we obtain the general integral formula[3]

$$tr\,\langle\tau_1|\,\tau_2\rangle_G{}^{QP} = \int d[q,\,p]e^{iW[q,\,p]}$$

Here the action functional $W[q, p]$ is

$$W[q,p] = \int_{\tau_2}^{\tau_1}[{}^1\!/_2(pdq - dpq) + d\tau(q(\tau)P(\tau) - p(\tau)Q(\tau)) + d\tau G(q(\tau)p(\tau)\tau)],$$

which is formed in essentially the same way as the action operator W_{12}, the multiplication order of noncommutative operator factors in G being replaced by suitable infinitesimal displacements of the parameter τ. The Hermitian operator G can always be constructed from symmetrized products of Hermitian functions of q and of p, and the corresponding numerical function is real. Thus the operator ${}^1\!/_2\{f_1(q(\tau),\, f_2(p(\tau)\}$ is represented by ${}^1\!/_2(f_1(q(\tau + \epsilon)) + f_1(q(\tau - \epsilon)))f_2(p(\tau))$, for example. One will expect to find that this averaged limit, $\epsilon \rightarrow 0$, implies no more than the direct use of $f_1(q(\tau)f_2(p(\tau))$, although the same statement is certainly not true of either term containing ϵ. Incidentally, it is quite sufficient to construct the action from pdq, for example, rather than the more symmetrical version, in virtue of the periodicity.

As a specialization of this trace formula, we place $Q = P = 0$ and consider the class of operators G that do not depend explicitly upon τ. Now we are computing

$$tr\, e^{iTG} = \int d[q, p] e^{iW[q, p]}$$

$$W[q, p] = \int_0^T (pdq + d\tau G),$$

in which we have used the possibility of setting $\tau_2 = 0$. A simple example is provided by $G = {}^1/_2(p^2 + q^2)$, where

$$W = {}^1/_2 T(p_0{}^2 + q_0{}^2) + \sum_{-\infty}^{\infty}{}' \left[p_n q_n + \frac{T}{4\pi|n|} (p_n{}^2 + q_n{}^2) \right]$$

and

$$tr\, e^{iT1/2(p^2+q^2)} = \int_{-\infty}^{\infty} \frac{dqdp}{2\pi} e^{iT1/2(p^2+q^2)} \times \prod_1^{\infty} \left[\int_{-\infty}^{\infty} \frac{dqdp}{2\pi} e^{1/2i(2qp + T/2\pi n(p^2+q^2))} \right]^2 =$$

$$\frac{i}{T} \prod_1^{\infty} \frac{1}{1 - \left(\dfrac{T}{2\pi n}\right)^2} = \frac{i}{2 \sin {}^1/_2 T}.$$

This example also illustrates the class of Hermitian operators with spectra that are bounded below and for which the trace of exp (iTG) continues to exist on giving T a positive imaginary component, including the substitution $T \rightarrow i\beta$, $\beta > 0$. The trace formula can be restated for the latter situation on remarking that the Fourier series depend only upon the variable $(\tau - \tau_2)/T = \lambda$, which varies from 0 to 1, and therefore

$$tr\, e^{-\beta G} = \int d[q, p] e^{-w[q, p]}$$

$$w[q, p] = \int_0^1 d\lambda \left[-ip \frac{dq}{d\lambda} + \beta G(q(\lambda)p(\lambda)) \right].$$

Another property of the example is that the contributions to the trace of all Fourier coefficients except $n = 0$ tend to unity for sufficiently small T or β. This is also true for a class of operators of the form $G = {}^1/_2 p^2 + f(q)$. By a suitable translation of the Fourier coefficients for $p(\lambda)$ we can write $w[q, p]$ as

$$w = \int_0^1 d\lambda \left[{}^1/_2 \beta p(\lambda)^2 + \frac{1}{2\beta} \left(\frac{dq}{d\lambda}\right)^2 + \beta f(q(\lambda)) \right].$$

For sufficiently small β, the term involving $dq/d\lambda$, to which all the Fourier coefficients of $q(\lambda)$ contribute except q_0, will effectively suppress these Fourier coefficients provided appropriate restrictions are imposed concerning singular points in the neighborhood of which $f(q)$ acquires arge negative values. Then $f(q(\lambda) \sim f(q_0))$ and we can reduce the integrations to just the contribution of q_0 and p_0, as expressed by

$$tr\, e^{-\beta G(q, p)} \sim \int_{-\infty}^{\infty} \frac{dqdp}{2\pi} e^{-\beta G(q, p)}.$$

Comparison with the previously obtained trace formula involving the ordering of operators shows that the noncommutativity of q and p is not significant in this limit. Thus we have entered the classical domain, where the incompatibility of physical properties at the microscopic level is no longer detectible. Incidentally, a first correction to the classical trace evaluation, stated explicitly for one degree of freedom, is

$$tr\, e^{-\beta[1/2p^2+f(q)]} \sim \int_{-\infty}^{\infty} \frac{dq}{(2\pi\beta)^{1/2}}\, e^{-\beta f(q)} \cdot \frac{{}^1/{}_2\beta(f''(q))^{1/2}}{\sinh {}^1/{}_2\beta(f''(q))^{1/2}},$$

which gives the exact value when $f(q)$ is a positive multiple of q^2.

Another treatment of the general problem can be given on remarking that the equations of motion implied by the stationary action principle,

$$\frac{dq}{d\tau} + \frac{\partial G}{\partial p} = Q, \quad \frac{dp}{d\tau} - \frac{\partial G}{\partial q} = P,$$

can be represented by functional differential equations[4]

$$\left[\frac{d}{d\tau}(-i)\frac{\delta}{\delta P(\tau)} + \frac{\partial G}{\partial p}\left(-i\frac{\delta}{\delta P}, i\frac{\delta}{\delta Q}\right) - Q(\tau)\right]\langle\tau_1|\,\tau_2\rangle_G^{QP} = 0$$

$$\left[\frac{d}{d\tau}\, i\frac{\delta}{\delta Q(\tau)} - \frac{\partial G}{\partial q}\left(-i\frac{\delta}{\delta P}, i\frac{\delta}{\delta Q}\right) - P(\tau)\right]\langle\tau_1|\,\tau_2\rangle_G^{QP} = 0.$$

These equations are valid for any such transformation function. The trace is specifically distinguished by the property of periodicity,

$$\left(\frac{\delta}{\delta Q(\tau_1)} - \frac{\delta}{\delta Q(\tau_2)}\right) tr\,\langle\tau_1|\,\tau_2\rangle_G^{QP} = \left(\frac{\delta}{\delta P(\tau_1)} - \frac{\delta}{\delta P(\tau_2)}\right)(tr) = 0$$

which asserts that the trace depends upon $Q(\tau)$, $P(\tau)$ only through the Fourier coefficients Q_n, P_n, and that the functional derivatives can be interpreted by means of ordinary derivatives:

$$\frac{\delta}{\delta Q(\tau)} = \frac{\partial}{\partial Q_0} + \sum_1^{\infty} (\pi n)^{-1/2}\left[-\sin\frac{2\pi n}{T}(\tau - \tau_2)\frac{\partial}{\partial Q_n} + \cos\frac{2\pi n}{T}(\tau - \tau_2)\frac{\partial}{\partial Q_{-n}}\right]$$

$$\frac{\delta}{\delta P(\tau)} = \frac{\partial}{\partial P_0} + \sum_1^{\infty} (\pi n)^{-1/2}\left[\cos\frac{2\pi n}{T}(\tau - \tau_2)\frac{\partial}{\partial P_n} + \sin\frac{2\pi n}{T}(\tau - \tau_2)\frac{\partial}{\partial P_{-n}}\right].$$

Now introduce the functional differential operator that is derived from the numerical action function $W_G[q, p]$ which refers only to the transformation generated by G, namely

$$W_G\left[-i\frac{\delta}{\delta P}, i\frac{\delta}{\delta Q}\right] = \int_{\tau_2}^{\tau_1}\left[\frac{\delta}{\delta Q}\, d\, \frac{\delta}{\delta P} + d\tau G\left(-i\frac{\delta}{\delta P}, i\frac{\delta}{\delta Q}, \tau\right)\right],$$

and observe that the differential equations are given by

$$(i[W_G, Q(\tau)] + Q(\tau))\; tr\,\langle\tau_1|\,\tau_2\rangle_G^{QP} = 0$$

$$(i[W_G, P(\tau)] + P(\tau))\; tr\,\langle\tau_1|\,\tau_2\rangle_G^{QP} = 0,$$

or by

$$e^{iW_G}Q(\tau)e^{-iW_G}(tr) = e^{iW_G}P(\tau)e^{-iW_G}(tr) = 0.$$

The latter form follows from the general expansion

$$e^A B e^{-A} = B + [A, B] + {}^1/_2![A, [A, B]] + \ldots$$

on noting that $[W_G, Q(\tau)]$, for example, is constructed entirely from differential operators and is commutative with the differential operator W_G. Accordingly,

$$Q(\tau)e^{-iW_G}(tr) = P(\tau)e^{-iW_G}(tr) = 0,$$

which asserts that $\exp(-iW_G)(tr)$ vanishes when multiplied by any of the Fourier coefficients Q_n, P_n and therefore contains a delta function factor for each of these variables. We conclude that

$$tr\,\langle\tau_1|\,\tau_2\rangle_G{}^{QP} = e^{iW_G[-i\,\delta/\delta P,\ i\,\delta/\delta Q]}\,\delta[Q, P]$$

where, anticipating the proper normalization constants,

$$\delta[Q, P] = \prod_{-\infty}^{\infty} 2\pi\delta(Q_n)\delta(P_n).$$

A verification of these factors can be given by placing $G = 0$, which returns us to the consideration of the special canonical transformations. In this procedure we encounter the typical term

$$e^{i\,\partial/\partial Q\,\partial/\partial P}\,2\pi\delta(Q)\delta(P) = e^{iQP}$$

the proof of which follows from the remarks that

$$\left(Q + i\frac{\partial}{\partial P}\right)e^{i\,\partial/\partial Q\,\partial/\partial P} = e^{i\,\partial/\partial Q\,\partial/\partial P}\,Q$$

and

$$\int_{-\infty}^{\infty}\frac{dP}{2\pi}\,e^{i\,\partial/\partial Q\,\partial/\partial P}\,2\pi\delta(Q)\delta(P) = \delta(Q).$$

The result is just the known form of the transformation function trace

$$tr\,\langle\tau_1|\,\tau_2\rangle^{QP} = 2\pi\delta(Q_0)\delta(P_0)e^{i-\sum_{\infty}^{\infty}{}' Q_nP_n}$$

When integral representations are inserted for each of the delta function factors in $\delta[Q, P]$, we obtain

$$\delta[Q, P] = \prod_{-\infty}^{\infty}\int_{-\infty}^{\infty}\frac{dq_n dp_n}{2\pi}\,e^{i(q_nP_n - p_nQ_n)} = \int d[q, p]\,e^{i\int_{\tau_2}^{\tau_1}d\tau(q(\tau)P(\tau) - p(\tau)Q(\tau))}$$

and the consequence of performing the differentiations in W_G under the integration signs is[5]

$$tr\,\langle\tau_1|\,\tau_2\rangle_G{}^{QP} = \int d[q, p]\,e^{iW[q,\,p]}$$

where the action function $W[q, p]$ now includes the special canonical transformation described by Q and P.

We shall also write this general integral formula as

$$tr\,\langle\tau_1|\,\tau_2\rangle_G{}^{QP} = \int d[q,\,p]\,e^{i\int_{\tau_2}^{\tau_1}d\tau(qP-pQ)}\,e^{iW_G[q,\,p]}$$

in order to emphasize the reciprocity between the trace, as a function of Q_n, P_n or functional of $Q(\tau)$, $P(\tau)$ and $\exp(iW_G[q,\,p])$ as a function of q_n, p_n or functional of $q(\tau)$, $p(\tau)$. Indeed,

$$e^{iW_G[q,\,p]} = \int d[Q,\,P]e^{-i\int_{\tau_2}^{\tau_1}d\tau(qP-pQ)}\,tr\,\langle\tau_1|\,\tau_2\rangle_G{}^{QP},$$

where

$$d[Q,\,P] = \prod_{-\infty}^{\infty}\frac{dQ_n dP_n}{2\pi}$$

is such that

$$\int d[Q,\,P]\delta[Q,\,P] = 1$$

A verification of the reciprocal formula follows from the latter property on inserting for the trace the formal differential operator construction involving $\delta[Q,\,P]$. The reality of $W_G[q,\,p]$ now implies that

$$\int d[Q,\,P]d[Q',\,P']e^{i\int d\tau(q(P-P')-p(Q-Q'))}\,(tr)^{QP*}\,(tr)^{Q'P'} = 1$$

or, equivalently,

$$\int d[Q,\,P](tr)^{X+X_1*}(tr)^{X+X_2} = \delta[Q_1 - Q_2,\,P_1 - P_2]$$

where X combines Q and P.

The trace possesses the composition property

$$tr\,\langle\tau_0|\,\tau_1\rangle_G{}^{QP} \times tr\,\langle\tau_1|\,\tau_2\rangle_G{}^{QP} = tr\,\langle\tau_0|\,\tau_2\rangle_G{}^{QP}.$$

The operation involved is the replacement of $Q(\tau)$, $P(\tau)$ in the respective factors by $Q(\tau) \mp q'\delta(\tau - \tau_1)$, $P(\tau) \mp p'\delta(\tau - \tau_1)$ followed by integration with respect to $dq'dp'/2\pi$. The explicit form of the left side is therefore

$$\int_{-\infty}^{\infty}\frac{dq'dp'}{2\pi}\,tr\,\langle\tau_0|\,e^{i[p(\tau_1)q'-p'q(\tau_1)]}\,|\,\tau_1\rangle_G{}^{QP}\,tr\,\langle\tau_1|\,e^{-i[p(\tau_1)q'-p'q(\tau_1)]}\,|\,\tau_2\rangle_G{}^{QP}$$

or

$$\sum_{a'a''}\int_{-\infty}^{\infty}\frac{dq'dp'}{2\pi}\,\langle a'\tau_0|\,U(q'p')\,|\,a'\tau_1\rangle\,\langle a''\tau_1|\,U(q'p')^\dagger\,|\,a''\tau_2\rangle =$$

$$\sum_{a'a''}\langle a'\tau_0|\,\delta(a',\,a'')\,|\,a''\tau_2\rangle = tr\,\langle\tau_0|\,\tau_2\rangle_G{}^{QP},$$

in view of the completeness of the operator basis formed from $q(\tau_1)$, $p(\tau_1)$.

No special relation has been assumed among τ_0, τ_1 and τ_2. If τ_0 and τ_2 are equated, one transformation function in the product is the complex conjugate of the other. We must do more than this, however, to get a useful result. The most general procedure would be to choose the special canonical transformation in $tr\,\langle\tau_2|\,\tau_1\rangle_G{}^{QP} = tr\,\langle\tau_1|\,\tau_2\rangle_G{}^{QP*}$ arbitrarily different from that in $tr\,\langle\tau_1|\,\tau_2\rangle_G{}^{QP}$. The calculational advantages that appear in this way will be explored elsewhere.

Here we shall be content to make the special canonical transformations differ only at τ_2. The corresponding theorem is

$$tr\,\langle\tau_2|\,\tau_1\rangle^{X+X'} \times tr\,\langle\tau_1|\,\tau_2\rangle^{X+X''} = 2\pi\delta(q'-q'')\delta(p'-p'')$$

where

$$Q'(\tau) = -q'\delta(\tau-\tau_2),\; P'(\tau) = -p'\delta(\tau-\tau_2)$$

and similarly for $Q''(\tau)$, $P''(\tau)$. This statement follows immediately from the orthonormality of the $U(q'p')$ operator basis on evaluating the left-hand side as

$$tr\, e^{-i[p(\tau_2)q'-p'q(\tau_2)]}\, e^{i[p(\tau_2)q''-p''q(\tau_2)]} = 2\pi\delta(q'-q'')\delta(p'-p'').$$

* Supported by the Air Force Office of Scientific Research (ARDC).

[1] These PROCEEDINGS, **46,** 883 (1960).

[2] These PROCEEDINGS, **46,** 570 (1960).

[3] This formulation is closely related to the algorithms of Feynman, *Phys. Rev.*, **84,** 108 (1951), *Rev. Mod. Phys.*, **20,** 36 (1948). It differs from the latter in the absence of ambiguity associated with noncommutative factors, but primarily in the measure that is used. See Footnote 5.

[4] These are directly useful as differential equations only when $G(qp)$ is a sufficiently simple algebraic function of q and p. The kinematical, group foundation for the representation of equations of motion by functional differential equations is to be contrasted with the dynamical language used in these PROCEEDINGS, **37,** 452 (1951).

[5] In this procedure, $q(\tau)$ and $p(\tau)$ are continuous functions of the parameter τ and the Fourier coefficients that represent them are a denumerably infinite set of integration variables. An alternative approach is the replacement of the continuous parameter τ by a discrete index while interpreting the derivative with respect to τ as a finite difference and constructing $\delta[Q, P]$ as a product of delta functions for each discrete τ value. With the latter, essentially the Feynman-Wiener formulation, the measure $d[q, p]$ is the product of $dq(\tau)dp(\tau)/2\pi$ for each value of τ, periodicity is explicitly imposed at the boundaries, and the limit is eventually taken of an infinitely fine partitioning of the interval $T = \tau_1 - \tau_2$. The second method is doubtless more intuitive, since it is also the result of directly compounding successive infinitesimal transformations but it is more awkward as a mathematical technique.

6.19 ADDENDUM: QUANTUM VARIABLES AND THE ACTION PRINCIPLE†

---

† Reproduced from the Proceedings of the National Academy of Sciences, Vol. 47, pp. 1075 -1083 (1961).

Reprinted from the Proceedings of the NATIONAL ACADEMY OF SCIENCES
Vol. 47, No. 7, pp. 1075–1083. July, 1961.

QUANTUM VARIABLES AND THE ACTION PRINCIPLE

BY JULIAN SCHWINGER

HARVARD UNIVERSITY AND UNIVERSITY OF CALIFORNIA AT LOS ANGELES

Communicated May 29, 1961

In previous communications, a classification of quantum degrees of freedom by a prime integer ν has been given,[1] and a quantum action principle has been constructed[2] for $\nu = \infty$. Can a quantum action principle be devised for other types of quantum variables? We shall examine this question for the simplest quantum degree of freedom, $\nu = 2$.

Let us consider first a single degree of freedom of this type. The operator basis is generated by the complementary pair of Hermitian operators

$$\xi_1 = 2^{-1/2}\sigma_1, \quad \xi_2 = 2^{-1/2}\sigma_2$$

which obey

$$\xi_k\xi_l + \xi_l\xi_k = \{\xi_k, \xi_l\} = \delta_{kl}.$$

The basis is completed by the unit operator and the product

$$\xi_3 = 2^{-1/2}\sigma_3 = -i2^{1/2}\xi_1\xi_2.$$

We have remarked[1] upon the well-known connection between the σ_k, $k = 1, 2, 3$, and three-dimensional rotations. In particular, the most general unitary operator that differs infinitesimally from unity is, apart from a phase factor, of the form

$$U(\delta\omega) = 1 + {}^1/_2 i\delta\omega\cdot\sigma,$$

and the corresponding operator transformation

$$\bar{\sigma} = U^{-1}\sigma U = \sigma - \delta\sigma$$

is the three-dimensional infinitesimal rotation

$$\delta\sigma = \delta\omega \times \sigma.$$

Accordingly, an infinitesimal transformation that varies ξ_1 and not ξ_2 can only be a rotation about the second axis,

$$\delta\xi_1 = \delta\omega_2\xi_3, \quad \delta\xi_2 = 0.$$

The corresponding infinitesimal generator is

$$G_1 = {}^1/_2\delta\omega_2\sigma_2 = i\xi_1\delta\xi_1 = -i\delta\xi_1\xi_1.$$

Similarly,

$$\delta\xi_1 = 0, \quad \delta\xi_2 = -\delta\omega_1\xi_3$$

is generated by

$$G_2 = i\xi_2\delta\xi_2 = -i\delta\xi_2\xi_2,$$

and the generator for the combination of these elementary transformations is ($k = 1, 2$)

$$G = i\sum_k \xi_k \delta\xi_k = -i\sum_k \delta\xi_k \xi_k.$$

It is evident that G_1 and G_2 must be supplemented by

$$G_3 = {}^1/_2 \delta\omega_3 \sigma_3$$

to form the infinitesimal generators of a unitary group, which is isomorphic to the three-dimensional rotation group. The transformation induced by G_3 is

$$\delta\xi_1 = -\delta\omega_3\xi_2, \quad \delta\xi_2 = \delta\omega_3\xi_1,$$

and

$$G_3 = {}^1/_2 i(\xi_1\delta\xi_1 + \xi_2\delta\xi_2).$$

Thus, the concentration on the complementary pair of operators ξ_1 and ξ_2 does not give a symmetrical expression to the underlying three-dimensional rotation group. This is rectified somewhat by using, for those special transformations in which ξ_1 and ξ_2 are changed independently, the generator of the variations ${}^1/_2\delta\xi_k$, $k = 1, 2$,

$$G_\xi = {}^1/_2 i\sum_k \xi_k \delta\xi_k$$

An arbitrary infinitesimal unitary transformation is described by the transformation function

$$\langle \tau + d\tau | \tau \rangle = \langle | [1 + id\tau G(\xi, \tau)] | \rangle.$$

Infinitesimal variations in τ, $\tau + d\tau$, and the structure of G induce

$$\delta''\langle \tau + d\tau | \tau\rangle = i\langle | \delta''[d\tau G(\xi, \tau)]\rangle = i\langle \tau + d\tau | \delta''[d\tau G(\xi(\tau), \tau)] | \tau\rangle.$$

To this we add δ', the transformations generated by G_ξ, performed independently but continuously in τ, on the states $\langle \tau + d\tau|$ and $|\tau\rangle$,

$$\delta'\langle \tau + d\tau | \tau\rangle = i\langle \tau + d\tau | [G_\xi(\tau + d\tau) - G_\xi(\tau)] | \tau\rangle.$$

Here, $G_\xi(\tau + d\tau) - G_\xi(\tau) = -{}^1/_2 i\sum \delta\xi_k(\tau + d\tau)\xi_k(\tau + d\tau) - {}^1/_2 i\sum \xi_k(\tau)\delta\xi_k(\tau)$

$$= -{}^1/_2 i\sum [\delta\xi_k(\tau + d\tau)\xi_k(\tau) + \xi_k(\tau + d\tau)\delta\xi_k(\tau)] + {}^1/_2\sum [\delta\xi_k, [\xi_k, d\tau G]],$$

or

$$G_\xi(\tau + d\tau) - G_\xi(\tau) = \delta'[-{}^1/_2 i\sum \xi_k(\tau + d\tau)\xi_k(\tau) + d\tau G(\xi(\tau), \tau)] - {}^1/_2\sum [\xi_k, [d\tau G, \delta\xi_k]],$$

where δ', in its effect upon operators, refers to the special variations $\delta\xi_k$, $k = 1, 2$, performed independently but continuously at τ and $\tau + d\tau$.

It is only if the last term is zero that one obtains the quantum action principle $[\delta = \delta' + \delta'']$

$$\delta\langle \tau + d\tau | \tau\rangle = i\langle \tau + d\tau | \delta[W] | \tau\rangle,$$

with

$$W(\tau + d\tau, \tau) = -{}^1/_2 i\sum_k \xi_k(\tau + d\tau)\xi_k(\tau) + d\tau G(\xi(\tau), \tau).$$

Since the special variation is such that $\delta\xi_1$ and $\delta\xi_2$ are arbitrary multiples of ξ_3, it is necessary that $[G, \xi_3]$ commute with ξ_1 and ξ_2. Hence this commutator must be a multiple of the unit operator, which multiple can only be zero, since the trace of the commutator vanishes or, alternately, as required by $[G, \xi_3^2] = 0$. For an action principle formulation to be feasible, it is thus necessary and sufficient that

$$[G, \xi_1\xi_2] = 0.$$

Terms in ξ_1 and ξ_2 are thereby excluded from G, which restriction is also conveyed by the statement that a permissible G must be an even function of the ξ_k, $k = 1, 2$. Apart from multiples of the unit operator, generating phase transformations, the only allowed generator is $\xi_1\xi_2$, which geometrically is a rotation about the third axis.

It should be noted that the class of variations δ' can be extended to include the one generated by G_3, without reference to the structure of G. Thus,

$$G_3(\tau + d\tau) - G_3(\tau) = -{}^1\!/_2 i\sum[\delta\xi_k(\tau + d\tau)\xi_k(\tau) + \xi_k(\tau + d\tau)\delta\xi_k(\tau)] + {}^1\!/_2\sum[\delta\xi_k, [\xi_k, d\tau G]],$$

where the latter term equals

$$ {}^1\!/_2\delta\omega_3[\xi_1, [\xi_2, d\tau G]] - {}^1\!/_2\delta\omega_3[\xi_2, [\xi_1, d\tau G]] = -{}^1\!/_2\delta\omega_3[d\tau G, [\xi_1, \xi_2]] = 1/i[d\tau G, G_3],$$

and therefore, for this kind of δ' variation,

$$G_3(\tau + d\tau) - G_3(\tau) = \delta'[-{}^1\!/_2 i\sum\xi_k(\tau + d\tau)\xi_k(\tau) + d\tau G].$$

The action operator for a finite unitary transformation is

$$W_{12} = \int_{\tau_2}^{\tau_1} d\tau\left[-{}^1\!/_2 i\sum\frac{d\xi_k}{d\tau}\xi_k + G\right]$$

or, more symmetrically,

$$W_{12} = \int_{\tau_2}^{\tau_1} d\tau\left[{}^1\!/_4 i\sum_k\left(\xi_k\frac{d\xi_k}{d\tau} - \frac{d\xi_k}{d\tau}\xi_k\right) + G\right],$$

since W_{12} is only defined to within an additive constant, and

$$\frac{d}{d\tau}(\xi_k(\tau))^2 = 0.$$

The principle of stationary action

$$\delta[W_{12}] = G_1 - G_2,$$

which refers to a fixed form of the operator $G(\xi(\tau), \tau)$, expresses the requirement that a finite transformation emerge from the succession of infinitesimal transformations. It will be instructive to see how the properties of the quantum variables are conversely implied by this principle. The discussion will be given without explicit reference to the single pair of variables associated with one degree of freedom, since it is of greater generality.

The bilinear concomitant

$$\delta({}^1\!/_4 i\sum[\xi_k, d\xi_k] + Gd\tau) - d({}^1\!/_4 i\sum[\xi_k, \delta\xi_k] + G\delta\tau) = {}^1\!/_2 i\sum[\delta\xi_k, d\xi_k] + \delta G d\tau - dG\delta\tau$$

shows that

$$\delta G = \frac{dG}{d\tau}\,\delta\tau - {}^1/_2 i\sum\left[\delta\xi_k, \frac{d\xi_k}{d\tau}\right]$$

and gives

$$G_{1,\,2} = {}^1/_4 i\sum[\xi_k, \delta\xi_k] + G\delta\tau\Big|_{\tau_1,\,\tau_2}.$$

The generator term $G\delta\tau$ evidently restates the transformation significance of the operator G. The effect upon operators is conveyed by the infinitesimal unitary transformation

$$\bar{F}(\xi(\tau), \tau) = (1 - iG\delta\tau)F(\xi(\tau), \tau)(1 + iG\delta\tau) = F(\xi(\tau + \delta\tau), \tau),$$

which is the general equation of motion,

$$\frac{dF}{d\tau} = \frac{\partial F}{\partial\tau} - \frac{1}{i}\,[F, G].$$

Let us take $\delta\xi_k$ to be a special variation, which we characterize by the following properties: (1) Each $\delta\xi_k$ anticommutes with every ξ_l,

$$\{\delta\xi_k, \xi_l\} = 0;$$

(2) the $\delta\xi_k(\tau)$ have no implicit τ dependence; and (3) every $\delta\xi_k$ is an arbitrary infinitesimal numerical multiple of a common nonsingular operator, which does not vary with τ. The second basic property asserts that

$$[\delta\xi_k, G] = 0$$

which restricts G to be an even function of the ξ_k, in virtue of the anticommutativity of the special variations with each member of this set. Furthermore, the τ derivatives of the special variations are also anticommutative with the ξ_k, according to property (3), and therefore,

$$\sum {}^1/_2\left[\delta\xi_k, \frac{d\xi_k}{d\tau}\right] = \frac{d}{d\tau}\sum {}^1/_2[\delta\xi_k, \xi_k] - \sum {}^1/_2\left[\frac{\partial\delta\xi_k}{\partial\tau}, \xi_k\right] =$$
$$\frac{d}{d\tau}\sum \delta\xi_k\xi_k - \sum\frac{\partial\delta\xi_k}{\partial\tau}\,\xi_k = \sum \delta\xi_k\,\frac{d\xi_k}{d\tau} = -\sum\frac{d\xi_k}{d\tau}\,\delta\xi_k.$$

Then, if we write

$$\delta\,G - \frac{\partial G}{\partial\tau}\,\delta\tau = \sum \delta\xi_k\,\frac{\partial_l G}{\partial\xi_k} = \sum\frac{\partial_r G}{\partial\xi_k}\,\delta\xi_k,$$

which defines the left and right derivatives of G with respect to the ξ_k, we get

$$\frac{dG}{d\tau} = \frac{\partial G}{\partial\tau},$$

which is consistent with the general equation of motion, and

$$\sum\delta\xi_k\left(i\,\frac{d\xi_k}{d\tau} + \frac{\partial_l G}{\partial\xi_k}\right) = \sum\left(-i\,\frac{d\xi_k}{d\tau} + \frac{\partial_r G}{\partial\xi_k}\right)\delta\xi_k = 0.$$

The nonsingular operator contained in every special variation can be cancelled from the latter equation, and the arbitrary numerical factor in each $\delta\xi_k(\tau)$ implies that

$$i\frac{d\xi_k}{d\tau} = \frac{\partial_r G}{\partial \xi_k} = -\frac{\partial_l G}{\partial \xi_k}.$$

In the opposite signs of left and right derivatives, we recognize the even property of the function G.

On comparing the two forms of the equation of motion for ξ_k we see that

$$[G, \xi_k] = \frac{\partial_r G}{\partial \xi_k}.$$

If we reintroduce the special variations, this reads

$$\frac{1}{i}[G, {}^1/_2\, i \sum \xi_k \delta\xi_k] = \sum \frac{\partial_r G}{\partial \xi_k}\, {}^1/_2 \delta\xi_k.$$

The left side is just the change induced in G by G_ξ, the generator of the special variations, which also appears in $G_{1,\,2}$, while the right side gives the result in terms of changes of the ξ_k by ${}^1/_2\delta\xi_k$. Since G is an arbitrary even function of the ξ_k, we accept this as the general interpretation of the transformation generated by G_ξ, which then asserts of any operator F that

$$\frac{1}{i}[F, G_\xi] = \sum \frac{\partial_r F}{\partial \xi_k}\, {}^1/_2\delta\xi_k = \sum {}^1/_2\delta\xi_k \frac{\partial_l F}{\partial \xi_k}.$$

The implication for an odd function of the ξ_k is

$$\{F, \xi_k\} = \frac{\partial_r F}{\partial \xi_k} = \frac{\partial_l F}{\partial \xi_k},$$

and the particular choice $F = \xi_l$ gives the basic operator properties of these quantum variables

$$\{\xi_k, \xi_l\} = \delta_{kl}.$$

In this way, we verify that the quantum action principle gives a consistent account of all the characteristics of the given type of quantum variable.

The operator basis of a single degree of freedom is used in a different manner when the object of study is the three-dimensional rotation group rather than transformations of the pair of complementary physical properties. The generator of the infinitesimal rotation $\delta\sigma = \delta\omega \times \sigma$ is

$$G_\omega = {}^1/_2\delta\omega\cdot\sigma = {}^1/_4 i\sigma\cdot\delta\sigma = -{}^1/_4 i\delta\sigma\cdot\sigma.$$

On applying this transformation independently but continuously to the states of the transformation function $\langle\tau + d\tau|\tau\rangle$, we encounter

$$G_\omega(\tau + d\tau) - G_\omega(\tau) = -{}^1/_4 i\delta\sigma(\tau + d\tau)\cdot\sigma(\tau) - {}^1/_4 i\sigma(\tau + d\tau)\cdot\delta\sigma(\tau) + {}^1/_4[\delta\sigma\cdot, [\sigma, d\tau G]].$$

The identity,

$$[\delta\sigma\cdot, [\sigma, d\tau G]] - [\sigma\cdot, [\delta\sigma, d\tau G]] = [d\tau G, [\sigma\cdot, \delta\sigma]],$$

combined with the equivalence of the two left-hand terms, both of which equal

$$\delta\omega\cdot\sigma\times,\ [\sigma,\ d\tau G],$$

then gives

$$G_\omega(\tau + d\tau) - G_\omega(\tau) = \delta'[-{}^1/_4 i\sigma(\tau + d\tau)\cdot\sigma(\tau) + d\tau G(\sigma(\tau),\ \tau)],$$

where δ' describes the independent operator variations $\delta\sigma(\tau) = \delta\omega(\tau) \times \sigma(\tau)$ at τ and $\tau + d\tau$. If we add the effect of independent variations of τ, $\tau + d\tau$, and the structure of G $[\delta'']$, we obtain an action principle[3] without restrictions on the form of G.

The action operator for a finite transformation is

$$W_{12} = \int_{\tau_2}^{\tau_1} d\tau \left[-{}^1/_4 i\frac{d\sigma}{d\tau}\cdot\sigma + G\right]$$

$$= \int_{\tau_2}^{\tau_1} d\tau \left[{}^1/_8 i\left(\sigma\cdot\frac{d\sigma}{d\tau} - \frac{d\sigma}{d\tau}\cdot\sigma\right) + G\right].$$

The most general form for G, describing phase transformations and rotations, is

$$G = g(\tau) + {}^1/_2\sigma\cdot\frac{d\omega(\tau)}{d\tau}.$$

We shall be content to verify that the principle of stationary action reproduces the equations of motion that also follow directly from the significance of G,

$$\frac{d\sigma}{d\tau} = = -\frac{1}{i}[\sigma,\ G] = \sigma \times \frac{d\omega}{d\tau}.$$

The action principle asserts that

$$\frac{1}{4}i\left[\delta\sigma\cdot,\ \frac{d\sigma}{d\tau}\right] + \delta G - \frac{\partial G}{\partial\tau}\delta\tau = 0,$$

where the operator variations are arbitrary infinitesimal rotations, $\delta\sigma = \delta\omega \times \sigma$. Hence the equations of motion are

$$-{}^1/_2 i\left(\sigma \times \frac{d\sigma}{d\tau} + \frac{d\sigma}{d\tau} \times \sigma\right) = \sigma \times \frac{d\omega}{d\tau},$$

which appear in the anticipated form on remarking that the left-hand side of the latter equation equals

$$\frac{d}{d\tau}\frac{1}{2i}\sigma \times \sigma = \frac{d}{d\tau}\sigma.$$

The extension to n degrees of freedom of type $\nu = 2$ requires some discussion. At first sight, the procedure would seem to be straightforward. Operators associated with different degrees of freedom are commutative, and the infinitesimal generators of independent transformations are additive, which implies an action operator of the previous form with the summation extended over the n pairs of complementary variables. But we should also conclude that G must be an even function of the complementary variables associated with each degree of freedom separately, and this is an unnecessarily strong restriction.

To loosen the stringency of the condition for the validity of the action principle, we replace the relationship of commutativity between different degrees of freedom by one of anticommutativity. Let $\xi^{(\alpha)}{}_{1,2,3}$ be the operators associated with the α degree of freedom. We define

$$\xi_1 = \xi_1{}^{(1)}, \quad \xi_2 = \xi_2{}^{(1)}$$

$$\xi_{2k+1,\,2} = \prod_{\alpha=1}^{k} (2^{1/2}\xi_3{}^{(\alpha)})\xi_{1,\,2}{}^{(k+1)}, \qquad k = 1, 2, \ldots n-1$$

and this set of $2n$ Hermitian operators obeys

$$\{\xi_k, \xi_l\} = \delta_{kl}, \qquad k, l = 1 \ldots 2n.$$

The inverse construction is

$$\xi_{1,\,2}{}^{(k+1)} = (-2i)^k \xi_1\xi_2 \ldots \xi_{2k}\xi_{2k+1,\,2},$$

and in

$$\xi_{2n+1} = (-2i)^n \xi_1\xi_2 \ldots \xi_{2n}\cdot 2^{-1/2},$$

we have an operator that extends by one the set of anticommuting Hermitian operators with squares equal to $^1/_2$. In particular this operator anticommutes with every ξ_k, $k = 1 \ldots 2n$.

An infinitesimal transformation that alters only ξ_k must be such that

$$\{\delta\xi_k, \xi_l\} = 0 \qquad l = 1 \ldots 2n,$$

which identifies $\delta\xi_k$ as an infinitesimal numerical multiple of ξ_{2n+1}. We shall write

$$^1/_2\delta\xi_k = -\delta\omega_k\xi_{2n+1}$$

$$= \frac{1}{i}[\xi_k, G_\xi]$$

and the generator of all these special variations is

$$G_\xi = \sum_{k=1}^{2n} \delta\omega_k \frac{1}{i}\, \xi_k\xi_{2n+1}.$$

The latter can also be written as

$$G_\xi = {}^1/_2 i \sum_{k=1}^{2n} \xi_k\delta\xi_k = -{}^1/_2 i \sum \delta\xi_k\xi_k.$$

On forming the commutator of two such generators, we get

$$\frac{1}{i}\,[G_\xi{}^{(1)}, G_\xi{}^{(2)}] = \sum_{k,\,l} {}^1/_2(\delta^{(1)}\omega_k\delta^{(2)}\omega_l - \delta^{(1)}\omega_l\delta^{(2)}\omega_k)\xi_{kl},$$

where the Hermitian operators

$$\xi_{kl} = \frac{1}{2i}[\xi_k, \xi_l] = -\xi_{lk}$$

are $n(2n - 1)$ in number, and obey $(k \neq l)$

$$\xi_{kl}^2 = {}^1/_4.$$

The generators G_ξ and their commutators can thus be constructed from the basis provided by the $n(2n + 1)$ operators ξ_{kl}, with k and l ranging from 1 to $2n + 1$. And, since

$$1/i[\xi_{kl}, \xi_{pq}] = \delta_{lq}\xi_{kp} - \delta_{kq}\xi_{lp} + \delta_{kp}\xi_{lq} - \delta_{lp}\xi_{kq},$$

these operators are the generators of a unitary transformation group, which has the structure of the Euclidean rotation group in $2n + 1$ dimensions. For $n > 1$, the operators ξ_{kl}, $k, l = 1 \,..\, 2n + 1$ can also be combined with the linearly independent set

$$\xi_{0k} = 2^{-1/2}\xi_k, \qquad k = 1 \,..\, 2n + 1$$

to form the $(n + 1)(2n + 1)$ generators of a similar group associated with rotations in $2n + 2$ dimensions.

It should also be noted that

$$G_\omega = {}^1/_2 \sum_{k,\, l\, =\, 1}^{2n} \delta\omega_{kl}\xi_{kl}, \qquad \delta\omega_{kl} = -\delta\omega_{lk}$$

induces the linear transformation

$$\delta\xi_k = \frac{1}{i}[\xi_k, G_\omega] = -\sum_{l=1}^{2n} \delta\omega_{kl}\xi_l,$$

and one can write

$$G_\omega = {}^1/_{2i} \sum_{k=1}^{2n} \xi_k\delta\xi_k = -{}^1/_{2i} \sum \delta\xi_k\xi_k.$$

These generators have the structure of the rotation group in $2n$ dimensions.

The discussion of the change induced in a transformation function $\langle\tau + d\tau|\tau\rangle$ by the special variations, applied independently but continuously at τ and $\tau + d\tau$, proceeds as in the special example $n = 1$, and leads to

$$\sum_{k=1}^{2n} [\xi_k, [d\tau G, \delta\xi_k]] = 0$$

for the condition that permits an action principle formulation. Each special variation is proportional to the single operator ξ_{2n+1}. It is necessary, therefore, that $[G, \xi_{2n+1}]$ commute with every ξ_k, $k = 1 \ldots 2n$, or equivalently, with each complementary pair of operators $\xi_{1,\,2}^{(\alpha)}$, $\alpha = 1 \ldots n$. Such an operator can only be a multiple of the unit operator and that multiple must be zero. Hence the infinitesimal generators of transformations that can be described by an action principle must commute with ξ_{2n+1}. Considered as a function of the anticommutative operator set ξ_k, $k = 1 \ldots 2n$ which generates the 2^{2n} dimensional operator basis, an admissible operator $G(\xi)$ must be even. This single condition replaces the set of n conditions that appear when commutativity is the relationship between different degrees of freedom. Of course, if the class of transformations under consideration is such that G is also an even function of some even-dimensional subsets of the ξ_k, one can consistently adopt commutativity as the relationship between the various subsets. Incidentally, the commutativity of G with $\xi_{2n+1}(\tau)$ asserts that the latter operator does not vary with τ. Accordingly, the special variations $\delta\xi_k(\tau)$ have no implicit operator dependence upon τ and one concludes that the special variations are anticommutative with the $2n$ fundamental variables ξ_k, without reference to the associated τ values.

The action principle is also valid for the linear variations induced by G_ω, without regard to the structure of G. Thus,

$$G_\omega(\tau + d\tau) - G_\omega(\tau) = -{}^1/_2 i \sum [\delta\xi_k(\tau + d\tau)\xi_k(\tau) + \xi_k(\tau + d\tau)\delta\xi_k(\tau)] + {}^1/_2 \sum [\delta\xi_k, [\xi_k, d_\tau G]],$$

and the last term equals

$$-{}^1/_4 \sum_{k,\, l} \delta\omega_{kl}[\xi_l, [\xi_k, d_\tau G]] + {}^1/_4 \sum \delta\omega_{kl}[\xi_k, [\xi_l, d_\tau G]] = \frac{1}{i}[d_\tau G, G_\omega],$$

which gives

$$G_\omega(\tau + d\tau) - G_\omega(\tau) = \delta'[-{}^1/_2 i \sum_k \xi_k(\tau + d\tau)\xi_k(\tau) + d_\tau G]$$

for this type of δ' variation.

The action operator associated with a finite transformation generalizes the form already encountered for $n = 1$,

$$W_{12} = \int_{\tau_2}^{\tau_1} d\tau \left[\frac{1}{4} i \sum_{k=1}^{2n} \left(\xi_k \frac{d\xi_k}{d\tau} - \frac{d\xi_k}{d\tau}\xi_k\right) + G(\xi(\tau), \tau)\right],$$

and the earlier discussion can be transferred intact. It may be useful, however, to emphasize that the special variations anticommute, not only with each $\xi_k(\tau)$, but also with $d\xi_k/d\tau$, since this property is independent of τ. Then it follows directly from the implication of the principle of stationary action,

$$\delta G = \frac{dG}{d\tau}\delta\tau - \frac{1}{2} i \sum_k \left[\delta\xi_k, \frac{d\xi_k}{d\tau}\right],$$

that the equations of motion are

$$i\frac{d\xi_k}{d\tau} = \frac{\partial_r G}{\partial \xi_k} = -\frac{\partial_l G}{\partial \xi_k}, \qquad k = 1 \ldots 2n.$$

The action principle is still severely restricted in a practical sense, for the generators of the special variations cannot be included in G since the operators $\xi_k\xi_{2n+1}$ are odd functions of the $2n$ fundamental variables. It is for the purpose of circumventing this difficulty, and thereby of converting the action principle into an effective computation device, that we shall extend the number system by adjoining an exterior or Grassmann algebra.

[1] These PROCEEDINGS, **46**, 570 (1960).

[2] *Ibid.*, **46**, 883 (1960).

[3] The possibility of using the components of an angular momentum vector as variables in an action principle was pointed out to me by D. Volkov during the 1959 Conference on High Energy Physics held at Kiev, U.S.S.R.

CHAPTER SEVEN

CANONICAL TRANSFORMATION FUNCTIONS

7.1 ORDERED ACTION OPERATOR

According to the significance of the action operator $W(q, \bar{q}, t)$ that defines a canonical transformation at time t, infinitesimal alterations of the eigenvalues and of t produce a change in the canonical transformation function $\langle q't|\bar{q}'t\rangle$

given by

$$\delta \langle q't|\bar{q}'t\rangle = i\langle q't|\ \delta W(q\ ,\ \bar{q}\ ,\ t)\ |\bar{q}'t\rangle\ . \qquad (7.1)$$

For some transformations, the commutation properties of the q and $\bar{q}$ variables can be used to rearrange the operator δW so that the q's everywhere stand to the left of the $\bar{q}$'s . This ordered differential expression will be denoted by $\delta \mathcal{W}(q\ ;\ \bar{q}\ ,\ t)$ and

$$\delta W(q\ ,\ \bar{q}\ ,\ t) = \delta \mathcal{W}(q\ ;\ \bar{q}\ ,\ t)\ . \qquad (7.2)$$

From the manner of construction of the latter operator, the variables q and $\bar{q}$ act directly on their eigenvectors in (7.1) and this equation becomes

$$\delta \langle q't|\bar{q}'t\rangle = i\ \delta \mathcal{W}(q'\ ;\ \bar{q}'t)\ \langle q't|\bar{q}'t\rangle\ . \qquad (7.3)$$

Hence $\delta \mathcal{W}$ must be an exact differential, and integration yields

$$\langle q't|\bar{q}'t\rangle = e^{i\mathcal{W}(q'\ ;\ \bar{q}'\ ,\ t)} \qquad (7.4)$$

in which a multiplicative integration constant is incorporated additively in $\mathcal{W}$. This constant is fixed, in part, by the composition properties of the transformation function. It is to be emphasized that the ordered operator $\mathcal{W}$ does not equal W and, indeed, is not Hermitian, should W possess this property. We have in effect already illustrated the ordering method by the construction of the transformation function $\langle q'|p'\rangle$. For Hermitian variables of the first kind, for example,

$$W(q\ ,\ p) = \sum \tfrac{1}{2}\{q_k\ ,\ p_k\} \tag{7.5}$$

whereas

$$\begin{aligned} \mathcal{W}(q\ ;\ p) &= \sum q_k p_k + i\,\frac{n}{2}\log 2\pi \\ &= W(q\ ,\ p) + i\,\frac{n}{2}(\log 2\pi + 1)\ . \end{aligned} \tag{7.6}$$

7.2 INFINITESIMAL CANONICAL TRANSFORMATION FUNCTIONS

The transformation function for an arbitrary infinitesimal canonical transformation is easily constructed in this way, if one uses the action operator

$$W(p\ ,\ \bar{q}) = W(p\ ,\ q) + W(q\ ,\ \bar{q})$$
$$= -\sum_a p_a \cdot \bar{q}_a - G(\bar{q}\ ,\ p)\quad , \tag{7.7}$$

for

$$\delta W(p\ ,\ \bar{q}) = -\sum\left(\delta p_a\ \bar{q}_a + p_a\ \delta\bar{q}_a\right) - \delta G(\bar{q}\ ,\ p) \tag{7.8}$$

is brought into the ordered form by performing the required operation on the infinitesimal quantity δG with the aid of the known commutation relations between $\bar{q} = q$ and p . It is convenient to supplement G with a numerical factor λ so that we obtain a one-parameter family of transformations that includes the desired one $(\lambda = 1)$ and the identity transformation $(\lambda=0)$. Then

$$\delta(\lambda G) = \delta\lambda\ G + \lambda\ \delta G \tag{7.9}$$

and the ordering operation is to be applied to G as well as to δG . From the former we obtain an equivalent operator which we call $G(p\ ;\ \bar{q})$ and thus

$$\delta W = -\delta\left[\sum p_a\bar{q}_a\right] - \delta\lambda\ G(p\ ;\ \bar{q}) - \lambda\ \delta G\quad . \tag{7.10}$$

The integrability of the ordered operator δW now demands that the ordered version of δG be simply the variation of $G(p;\bar{q})$ and thus $(\lambda=1)$

$$W(p\,;\,\bar{q}) = -\sum p_a \bar{q}_a - G(p\,;\,\bar{q}) + \text{Const.} \tag{7.11}$$

in which the additive constant is that appropriate to the identity transformation, $W(p;q)$, and depends upon the kinds of variables employed. Hence

$$\langle p'|\bar{q}'\rangle = e^{iW(p'\,;\,\bar{q}')} = C\,e^{-i\sum p'_a q'_a - iG(p'\,;\,q')} , \tag{7.12}$$

or

$$\langle p'|\bar{q}\rangle = \langle p'|q'\rangle e^{-iG(p'\,;\,q')} , \tag{7.13}$$

and in view of the infinitesimal nature of G

$$\begin{aligned} \langle p'|\bar{q}'\rangle &= \langle p'|q'\rangle[1 - iG(p'\,;\,q')] \\ &= \langle p'|1 - iG|q'\rangle , \end{aligned} \tag{7.14}$$

which simply repeats the significance of G as the generator of the infinitesimal transformation. When the transformation corresponds to an infinitesimal change of parameters, (7.12) reads

$$\langle p'\tau|q'\ \tau-d\tau\rangle = Ce^{-i\sum p_a'q_a' + i\sum d\tau_r\, G_{(r)}(p';q',\tau)} \tag{7.15}$$

and, in particular,

$$\langle p't|q'\ t-dt\rangle = Ce^{-i\sum p_a'q_a' - i\,dt\, H(p';q',t)} \quad . \tag{7.16}$$

The composition properties of transformation functions can now be used to introduce other choices of canonical variables. For example, with Hermitian variables of the first kind,

$$\langle q'|\bar{q}''\rangle = \int \langle q'|p'\rangle\, dp'\, \langle p'|\bar{q}''\rangle \tag{7.17}$$

$$= \int \frac{dp'}{(2\pi)^n}\, e^{i\sum (q_k'-q_k'')p_k' - iG(p';q'')}$$

according to (4.21) and (7.12). The integrations are easily performed if G is a linear function of the p variables, or a non-singular quadratic function. In the first situation

$$G(p'\ ;\ q'') = G(0\ ;\ q'') + \sum p_k' \frac{\partial G(p';q'')}{\partial p_k'} \tag{7.18}$$

and

$$\langle q'|\bar{q}''\rangle = e^{-iG(0;q'')}\,\delta\left(q' - q'' - \frac{\partial G}{\partial p'}\right) \quad . \qquad (7.19)$$

For the quadratic functions of p, it is convenient to move the origin of the p' variables to the point defined by

$$q_k' - q_k'' = \frac{\partial G(p;q'')}{\partial p_k} \quad . \qquad (7.20)$$

Then

$$\langle q'|\bar{q}''\rangle = e^{iW} \int \frac{dp'}{(2\pi)^n} \exp\left(-\frac{i}{2}\sum p_k' p_\ell' \frac{\partial^2 G}{\partial p_k \partial p_\ell}\right)$$

$$= \left[(2\pi i)^n \det \frac{\partial^2 G}{\partial p \partial p}\right]^{-\frac{1}{2}} e^{iW} \qquad (7.21)$$

where

$$\begin{aligned} W &= \sum (q_k' - q_k'') p_k - G(p\,;\,q'') \\ &= \sum p_k \frac{\partial G(p\,;\,q'')}{\partial p_k} - G(p\,;\,q'') \quad . \end{aligned} \qquad (7.22)$$

On eliminating p with the aid of (7.20), we find

$$W = \tfrac{1}{2} \sum \left\{q_k' - q_k'' - \frac{\partial}{\partial p_k} G(0\,,\,q'')\right\} \qquad (7.23)$$

$$\times M_{k\ell}\left(q_\ell' - q_\ell'' - \frac{\partial}{\partial p_\ell} G(0\,,\,q'')\right) - G(0\,,\,q'')$$

in which $M(q'')$ is the matrix inverse to $\partial^2 G/\partial p \partial p$ In the limit as the latter approaches zero the delta function form (7.19) is obtained. Both results are also easily derived directly, without reference to the intermediate p representation. For the example of an infinitesimal Hamiltonian-Jacobi transformation in which H meets the requirements of a non-singular quadratic dependence on p, we have

$$\begin{aligned} \langle q't|q''\,t-dt\rangle &= \langle q'\,t+dt|q''t\rangle \\ &= \left[\det \frac{m}{2\pi i dt}\right]^{\frac{1}{2}} e^{iw} \end{aligned} \tag{7.24}$$

where m is the matrix inverse to $\partial^2 H/\partial p \partial p$ and

$$\begin{aligned} w = dt\, [\tfrac{1}{2} \sum &\left(\frac{q_k' - q_k''}{dt} - \frac{\partial}{\partial p_k} H(0;q'',t)\right) \\ &\times m_{k\ell} \left(\frac{q_\ell' - q_\ell''}{dt} - \frac{\partial}{\partial p_\ell} H(0;q'',t)\right) - H(0;q'',t)] \quad . \end{aligned} \tag{7.25}$$

An equivalent form for the latter is

$$\begin{aligned} w = dt\, [\tfrac{1}{2} \sum_{k\ell} &\frac{q_k' - q_k''}{dt} m_{k\ell} \frac{q_\ell' - q_\ell''}{dt} \\ &+ \sum_k \frac{q_k' - q_k''}{dt} p_{0k}(q'',t) - H(p_0;q'',t)] \end{aligned} \tag{7.26}$$

in which p_0 is defined by

$$\frac{\partial}{\partial p_{0k}} H(p_0;q'',t) = 0 \quad . \tag{7.27}$$

7.3 FINITE CANONICAL TRANSFORMATION FUNCTIONS

The transformation function for a finite canonical transformation can be constructed by the repeated composition of that for an infinitesimal transformation. To evaluate $\langle\tau_1|\tau_2\rangle$ along a particular path in the parameter space, we choose N intermediate points on the path, and compute

$$\langle\tau_1|\tau_2\rangle = \langle\tau_1|\tau^{(1)}\rangle \times \langle\tau^{(1)}|\tau^{(2)}\rangle \times \ldots \times \langle\tau^{(N)}|\tau_2\rangle \tag{7.28}$$

in which $\times$ signifies the application of a method of composition appropiate to the canonical variables employed. In the limit $N \to \infty$ the required path is traced as an infinite sequence of infinitesimal transformations. With Hermitian variables of the first kind and composition by integration, for example, we begin with

$$\langle q'\tau_1|q''\tau_2\rangle \tag{7.29}$$

$$= \int \frac{dp'}{(2\pi)^n} e^{i\sum (q_k'-q_k'')p_k' + i\sum d\tau_r G_{(r)}(p';q'',\tau)}$$

and obtain

$$\langle q'\tau_1|q''\tau_2\rangle = \operatorname*{Lim}_{N\to\infty} \int \prod_{j=1}^{N+1} \left[\frac{dq^{(j)}\, dp^{(j)}}{(2\pi)^n} \right.$$

$$\times \exp\{i \sum_k \left(q_k^{(j-1)} - q_k^{(j)}\right) p_k^{(j)} \tag{7.30}$$

$$\left. + i \sum_r \left(\tau_r^{(j-1)} - \tau_r^{(j)}\right) G_{(r)}\left(p^{(j)};q^{(j)},\tau^{(j)}\right) \}\right]$$

$$\times \delta\left(q^{(N+1)} - q''\right) \quad ,$$

in which

$$\tau^{(0)} = \tau_1 \quad , \quad \tau^{(N+1)} = \tau_2 \quad , \quad q^{(0)} = q' \tag{7.31}$$

and the delta function factor enforces the restriction $q^{(N+1)} = q''$. For the Hamilton-Jacobi transformation with a Hamiltonian that depends quadratically on p, this general form reduces to

$$\langle q't_1|q''t_2\rangle = \lim_{N\to\infty} \int \prod_{j=1}^{N+1} \Big[dq^{(j)}$$

$$\times \left(\det \left\{ \frac{1}{2\pi i} \frac{\partial^2 w}{\partial q^{(j-1)} \partial q^{(j-1)}} \right\} \right)^{\frac{1}{2}} \tag{7.32}$$

$$\times e^{iw\left(q^{(j-1)}t^{(j-1)}\,,\, q^{(j)}t^{(j)}\right)} \Big] \; \delta\left(q^{(N+1)} - q''\right)$$

in which w is given by (7.25) or (7.26). The generally applicable technique of composition by differentiation can be applied directly to the infinitesimal transformation function (7.15) and we obtain

$$\langle p'\tau_1|q'\tau_2\rangle$$

$$= \lim_{N\to\infty} \prod_{j=1}^{N+1} \left[\left\langle p^{(j-1)}\tau^{(j-1)}\right| i\,\frac{\partial_\ell}{\partial p^{(j)}}\;\tau^{(j)}\right\rangle\Bigg] \times \exp\left[-i\sum_a p_a^{(N+1)}\,q_a'\right]\Bigg|_{p^{(1)}=\,=\ldots p^{(N+1)}=0}$$

$$= \lim_{N\to\infty} \prod_{j=1}^{N+1} \exp\left[\sum_a p_a^{(j-1)}\,\frac{\partial_\ell}{\partial p_a^{(j)}} + i\sum_r \left(\tau_r^{(j-1)}-\tau_r^{(j)}\right) \times G_{(r)}\left(p^{(j-1)}\;;\; i\,\frac{\partial_\ell}{\partial p^{(j)}}\;,\;\tau^{(j)}\right)\right] \times \exp\left[-i\sum_a p_a^{(N+1)} q_a'\right]\Bigg|_{p^{(1)}=\ldots p^{(N+1)}=0} \quad (7.33)$$

apart from numerical factors of (2π) for Hermitian variables of the first kind. The boundary values are

$$\tau^{(0)} = \tau_1 \;,\quad \tau^{(N+1)} = \tau_2 \;,\quad p^{(0)} = p' \qquad (7.34)$$

and the sense of multiplication is that of (7.28).

7.4 ORDERED OPERATORS. THE USE OF CANONICAL TRANSFORMATION FUNCTIONS

On writing (7.15) as

$$\langle p'\tau|q'\ \tau-d\tau\rangle = \langle p'|q'\rangle\, e^{i\sum d\tau_r\, G_{(r)}(p';q',\tau)}$$

$$= \left\langle p'\right| e^{i\sum d\tau_r\, G_{(r)}(q,p,\tau)} \left| q'\right\rangle \tag{7.35}$$

the transformation function is obtained as a matrix in which the states and canonical variables do not depend on τ . The result of composing successive transformations then appears as the matrix of the product of the associated operators and thus

$$\langle\tau_1|\tau_2\rangle$$

$$= \lim_{N\to\infty} \left\langle \left| \prod_{j=1}^{N+1} \exp\left[i \sum_r \left(\tau_r^{(j-1)}-\tau_r^{(j)}\right) G_{(r)}\left(x,\tau^{(j)}\right)\right]\right|\right\rangle$$

$$= \left\langle \left| \left(\exp\left[i \int_{\tau_2}^{\tau_1} \sum d\tau_r\, G_{(r)}(x,\tau)\right] \right)_+ \right|\right\rangle \tag{7.36}$$

which defines the ordered exponential operator. The + subscript refers to the manner of multiplication, in which the positions of the operators correspond to the path in parameter space as deformed into a straight line, with τ_2 on the right and τ_1 on the left. We speak of negative rather than positive ordering if the sense of multiplication is reversed, so that

$$\left(\exp i \int_{\tau_2}^{\tau_1} \sum d\tau_r \, G_{(r)} \right)_+^{\dagger} = \left(\exp -i \int_{\tau_2}^{\tau_1} \sum d\tau_r \, G_{(r)} \right)_- . \tag{7.37}$$

When the generators are independent of τ, and the path is a straight line, the ordering is no longer significant and

$$\langle \tau_1 | \tau_2 \rangle = \left\langle \left| e^{i \sum \tau_r G_{(r)}} \right| \right\rangle \tag{7.38}$$

in analogy with the more specific group structure (6.39). Should the operators $G_{(r)}$ commute, thus generating an Abelian group, the exponential form (7.38) is valid without reference to the integration path. Let us suppose that the generators constitute, or can be extended to be a complete set of commuting Hermitian operators. Then a G representation exists, and a canonical transformation function describing the transformations of the group, say $\langle p'\tau_1|q'\tau_2\rangle$, can be exhibited as

$$\begin{aligned} \langle p'\tau_1|q'\tau_2\rangle &= \left\langle p'\middle| e^{i\sum \tau_r G_{(r)}} \middle| q'\right\rangle \\ &= \sum_{G'} \langle p'|G'\rangle \, e^{i\sum \tau_r G'_{(r)}} \langle G'|q'\rangle \quad , \end{aligned} \tag{7.39}$$

which shows how the canonical transformation function serves to determine the eigenvalues of the operators G as well as wave functions representing the G states.

7.5 AN EXAMPLE

An elementary example for $n = 1$ is provided by the Hermitian operator

$$G = -i\lambda pq = \lambda\, q^\dagger q \tag{7.40}$$

in which the non-Hermitian canonical variables may be of either type. A direct construction of the canonical transformation function can be obtained from (7.33) by repeated application of

$$\exp\left[(1 + i\lambda\, d\tau\,)p^{(j-1)} \frac{\partial_\ell}{\partial p^{(j)}}\right]$$

$$\times \exp\left[(1 + i\lambda\, d\tau\,)p^{(j)} \frac{\partial_\ell}{\partial p^{(j+1)}}\right]\Bigg|_{p^{(j)}=0} \tag{7.41}$$

$$= \exp\left[(1 + i\lambda\, d\tau)^2 p^{(j-1)} \frac{\partial_\ell}{\partial p^{(j+1)}}\right] ,$$

namely

$$\langle q^{\dagger\prime}\tau_1|q'\tau_2\rangle = \exp\,[e^{i\lambda\tau}\, q^{\dagger\prime}\; q']$$

$$= \sum_{n=0} \frac{(q^{\dagger\prime})^n}{(n!)^{\frac{1}{2}}}\, e^{in\lambda\tau}\, \frac{(q')^n}{(n!)^{\frac{1}{2}}} \; . \tag{7.42}$$

The eigenvalues thus obtained are

$$(q^{\dagger}q)' = n = 0\;,\;1\;,\ldots \tag{7.43}$$

and wave functions of these states are given by

$$\langle q^{\dagger\prime}|n\rangle = \frac{(q^{\dagger\prime})^n}{(n!)^{\frac{1}{2}}} \;, \qquad \langle n|q'\rangle = \frac{(q')^n}{(n!)^{\frac{1}{2}}} \;. \tag{7.44}$$

The derivation applies generally to both types of variables. However, for variables of the second kind $(q')^2 = (q^{\dagger\prime})^2 = 0$, and the summation in (7.42) terminates after $n = 0, 1$, which are indeed the only eigenvalues of $q_\alpha^\dagger q_\alpha$. The result (7.42) is reached more quickly, however, along the lines of (5.78). For non-Hermitian variables symmetrization or antisymmetrization $(\cdot)$ is unnecessary and it is omitted in (7.40). No ordering is required to obtain

$$W(q^\dagger ; q_0 , \tau) = - iq^\dagger e^{i\lambda\tau} q_0 \tag{7.45}$$

which immediately gives (7.42). An alternative procedure is

$$\begin{aligned}\left(\frac{1}{i}\right) \frac{\partial}{\partial\tau} \langle q^{\dagger\prime}\tau_1|q'\tau_2\rangle &= \langle q^{\dagger\prime}\tau_1|\lambda q^\dagger(\tau_1)q(\tau_1)|q'\tau_2\rangle \\ &= \lambda q^{\dagger\prime} e^{i\lambda\tau} q'\langle q^{\dagger\prime}\tau_1|q'\tau_2\rangle ,\end{aligned} \tag{7.46}$$

according to the solution of the equations of motion

$$q(\tau_1) = e^{i\lambda\tau} q(\tau_2) , \tag{7.47}$$

which gives (7.42) on integration when combined with the initial condition provided by

$$\langle q^{\dagger\prime} | q' \rangle = e^{q^{\dagger\prime} q'} \quad . \tag{7.48}$$

7.6 ORDERED OPERATORS AND PERTURBATION THEORY

A more general ordered operator form appears on decomposing the generators into two additive parts

$$G_{(r)} = G^0_{(r)} + G^1_{(r)} \tag{7.49}$$

Then (7.15) can be written

$$\langle p'\tau | q' \ \tau - d\tau \rangle_G \tag{7.50}$$

$$= \left\langle p'\tau \middle| q' \ \tau - d\tau \right\rangle_{G^0} \exp \left(i \sum d\tau_r \ G^1_{(r)} (p';q',\tau) \right)$$

$$= \left\langle p'\tau \right| \ \exp \left[i \sum d\tau_r \ G^1_{(r)} \left(x(\tau),\tau \right) \right] \ \left| q' \ \tau - d\tau \right\rangle_{G^0}$$

in which the subscripts indicate that the infinitesimal parameter change is governed by the generators G or G^0, respectively. The resulting finite transformation function is

$$\langle \tau_1 | \tau_2 \rangle_G \tag{7.51}$$

$$= \left\langle \tau_1 \right| \left(\exp \left[i \int_{\tau_2}^{\tau_1} \sum d\tau_r \ G^1_{(r)} \left(x(\tau),\tau \right) \right] \right)_+ \left| \tau_2 \right\rangle_{G^0}$$

in which the dynamical variables and states that appear on the right vary along the integration path in accordance with the generators $G^0_{(r)}$. For the example of the Hamilton-Jacobi transformation we have

$$\langle t_1|t_2\rangle_H \tag{7.52}$$

$$= \left\langle t_1\right| \left(\exp\left[-i \int_{t_2}^{t_1} dt\ H^1(x(t),t)\right] \right)_+ \left|t_2\right\rangle_{H^0} \quad .$$

Another derivation of the latter result proceeds directly from the fundamental dynamical principle, now applied to a change in the dynamical characteristics of the system. Thus, for a system with the Hamiltonian

$$H_\lambda = H^0 + \lambda H^1$$

an infintesimal change of the parameter λ produces a change in Hamiltonian, and hence of a canonical transformation function $\langle t_1|t_2\rangle$, described by

$$\frac{\partial}{\partial\lambda} \langle t_1|t_2\rangle = -\, i\left\langle t_1\right| \int_{t_2}^{t_1} dt\; H^1(x(t),t) \left|t_2\right\rangle$$

$$= -\, i \int_{t_2}^{t_1} dt\, \langle t_1|t\rangle \times \langle t|H^1(x(t),t)|t\rangle \times \langle t|t_2\rangle \quad . \tag{7.53}$$

In the latter form, the dependence upon λ appears in the two transformation functions $\langle t_1|t\rangle$ and $\langle t|t_2\rangle$. A second differentiation yields

$$\frac{\partial^2}{\partial\lambda^2} \langle t_1|t_2\rangle = (-i)^2 \tag{7.54}$$

$$\times \int_{t_2}^{t_1} dt \left[\int_{t}^{t_1} dt'\, \langle t_1|H^1(x(t'),t')H^1(x(t),t)|t_2\rangle \right.$$

$$\left. + \int_{t_2}^{t} dt'\, \langle t_1|H^1(x(t),t)H^1(x(t'),t')|t_2\rangle \right]$$

in which the two terms are equal, the limits of integration being alternative ways of expressing the ordered nature of the operator product. One can also write

$$\frac{\partial^2}{\partial\lambda^2} \langle t_1 | t_2 \rangle = (-i)^2$$

$$\times \int_{t_2}^{t_1} dt \int_{t_2}^{t_1} dt' \langle t_1 | \left(H^1(x(t),t) H^1(x(t'),t')\right)_+ | t_2 \rangle$$

$$= \left\langle t_1 \right| \left(-i \int_{t_2}^{t_1} dt\, H^1\right)_+^2 \left| t_2 \right\rangle \tag{7.55}$$

and the general statement is

$$\left(\frac{\partial}{\partial\lambda} \right)^n \langle t_1 | t_2 \rangle = \left\langle t_1 \right| \left(-i \int_{t_2}^{t_1} H^1\right)_+^n \left| t_2 \right\rangle . \tag{7.56}$$

Hence the transformation function for the system with Hamiltonian $H = H^0 + H^1$ $(\lambda=1)$ can be obtained formally as a power series expansion about $\lambda = 0$, and is thus expressed in terms of properties of the system with Hamiltonian H^0,

$$\left\langle t_1 \middle| t_2 \right\rangle_H = \left\langle t_1 \right| \sum_{n=0} \frac{1}{n!} \left(-i \int_{t_2}^{t_1} dt\, H^1\right)_+^n \left| t_2 \right\rangle_{H^0} . \tag{7.57}$$

The identification of this result with (7.52) supplies the expansion of the ordered exponential

$$\left(\exp \left[-i \int_{t_2}^{t_1} dt\, H^1(x(t),t)\right]\right)_+$$

$$= \sum_{n=0} \frac{1}{n!} (-i)^n \int_{t_2}^{t_1} dt^{(1)} \dots dt^{(n)}$$

$$\times \left(H^1(x(t^{(1)}),t^{(1)}) \dots H^1(x(t^{(n)}),t^{(n)})\right)_+ \tag{7.58}$$

$$= \sum_{n=0} (-i)^n \int_{t_2}^{t_1} dt^{(1)} \int_{t_2}^{t^{(1)}} dt^{(2)} \dots \int_{t_2}^{t^{(n-1)}} dt^{(n)}$$

$$\times H^1(x(t^{(1)}),t^{(1)}) \dots H^1(x(t^{(n)}),t^{(n)}) \quad .$$

The transformation function construction given in (7.52) is the foundation of perturbation theory, whereby the properties of a dynamical system are inferred from the known characteristics of another system. The expansion (7.58) is the basis of related approximation procedures.

7.7 USE OF THE SPECIAL CANONICAL GROUP

The properties of the special canonical group can be exploited as the foundation of a technique for obtaining canonical transformation functions. If we are interested in the Hamilton-Jacobi transformation, with infinitesimal generator $-H\,\delta t$ we consider the extended transformation characterized by

$$G = - H \, \delta t + \sum \left(p_a \, \delta q_a - \delta p_a \, q_a\right) \qquad (7.59)$$

which includes infinitesimal displacements of the canonical variables. On supposing such displacements to be performed independently within each infinitesimal time interval, as described by

$$\delta q_a(t) = - Q_a(t) \, \delta t \qquad \delta p_a(t) = - P_a(t) \, \delta t \quad , \qquad (7.60)$$

the infinitesimal generator becomes

$$G = - [H + \sum \left(p_a Q_a - P_a q_a\right)] \, \delta t \quad , \qquad (7.61)$$

and the extended transformation appears as a Hamilton-Jacobi transformation with an effective Hamiltonian. The corresponding equations of motions are

$$\frac{dq_a}{dt} = \frac{\partial_\ell H}{\partial \, p_a} + Q_a \quad , \qquad \frac{dp_a}{dt} = - \frac{\partial_r H}{\partial q_a} + P_a \qquad (7.62)$$

which exhibits the independent changes in the canonical variables that occur in a small time interval. If the displacements Q_a and P_a are localized at a time t_0 which means that

$$Q_a(t) = q_a' \, \delta(t-t_o) \quad ; \quad P_a(t) = p_a' \, \delta(t-t_a) \ , \tag{7.63}$$

the equations of motion imply a finite discontinuity in the canonical variables on passing through the time t_0 ,

$$\begin{aligned} q_a(t_0+0) - q_a(t_0-0) &= q_a' \ , \\ p_a(t_0+0) - p_a(t_0-0) &= p_a' \ . \end{aligned} \tag{7.64}$$

As an application of the latter result, the arbitrary eigenvalues of canonical variables that specify states at a certain time can be replaced by convenient standard values, provided the compensating canonical displacement is included. Thus with $t_1 > t_2$,

$$\langle p' t_1 | q' t_2 \rangle = \langle p'=0 \ , \ t_1+0 | q'=0 \ , \ t_2-0 \rangle_{QP} \tag{7.65}$$

where the displacements localized at the terminal times,

$$Q_a(t) = q_a' \, \delta(t-t_2) \ , \quad P_a(t) = - p_a' \, \delta(t-t_1) \tag{7.66}$$

do indeed convert the standard states into the desired ones at times $t_1(-0)$ and $t_2(+0)$. A proof of equivalence can also be obtained from

(7.52), applied to produce the transformation function of the system with Hamiltonian $H + \sum (p_a Q_a - P_a q_a)$ from that with Hamiltonian H,

$$\langle t_1 | t_2 \rangle_{HQP}$$

$$= \left\langle t_1 \right| \left(\exp \left[-i \int_{t_2}^{t_1} dt \sum_a (p_a Q_a - P_a q_a) \right] \right)_+ \left| t_2 \right\rangle_H . \tag{7.67}$$

With the localized displacement (7.66), the ordering of the exponential operator is immediate and

$$\langle p'=0\ t_1 | q'=0\ t_2 \rangle_{QP}$$

$$= \left\langle p'=0\ t_1 \right| e^{-i \sum p'_a q_a(t_1)} e^{-i \sum p_a(t_2) q'_a} \left| q'=0\ t_2 \right\rangle$$

$$= \langle p' t_1 | q' t_2 \rangle \tag{7.68}$$

since the two exponential operators produce the required canonical transformations at times t_1 and t_2.

7.8 VARIATIONAL DERIVATIVES

A more significant use of the special canonical group appears on considering arbitrary displacements throughout the interval between t_1 and t_2.

It is now convenient to unify the canonical variables and write

$$-\sum_a (p_a Q_a - P_a q_a) = \sum_k (P_k q_k - Q_k p_k) + \sum_\alpha (P_\alpha q_\alpha + Q_\alpha p_\alpha)$$
$$= \sum_a X_a x_a \quad . \tag{7.69}$$

An infinitesimal variation of $X_a(t)$ produces the corresponding transformation function change

$$\delta_X \langle t_1 | t_2 \rangle = i \left\langle t_1 \right| \int_{t_2}^{t_1} \sum \delta X_a(t)\; x_a(t) \left| t_2 \right\rangle$$
$$= \int_{t_2}^{t_1} dt \sum_a \delta X_a'(t) \frac{\delta_\ell}{\delta X_a'(t)} \langle t_1 | t_2 \rangle \tag{7.70}$$

which defines the left variational derivative of the transformation function with respect to $X_a'(t)$,

$$\frac{1}{i} \frac{\delta_\ell}{\delta X_a'(t)} \langle t_1 | t_2 \rangle = \langle t_1 | \rho x_a(t) | t_2 \rangle \quad . \tag{7.71}$$

The operator structure of the displacements that is used here,

$$X_a(t) = \rho X_a'(t) \tag{7.72}$$

refers explicitly to variables of the second kind where the X_a' are anticommuting exterior algebra elements and ρ is the operator that anticommutes with all dynamical variables, but it also covers variables of the first kind on replacing "anticommuting" by "commuting". In particular, $\rho \to 1$. The effect of a second variation is obtained from (7.55) as

$$\delta_x^2 \langle t_1|t_2\rangle = i^2 \int_{t_2}^{t_1} dt \int_{t_2}^{t_1} dt' \times \sum_{ab} \langle t_1| \left(\delta X_a(t)\ x_a(t)\ \delta X_b(t')\ x_b(t')\right)_+ |t_2\rangle$$

$$= \int_{t_2}^{t_1} dt\, dt' \times \sum_{ab} \delta X_b'(t')\ \delta X_a'(t)\ \frac{\delta_\ell}{\delta X_a'(t\ )}\ \frac{\delta_\ell}{\delta X_b'(t')} \langle t_1|t_2\rangle \quad , \tag{7.73}$$

and

$$\frac{1}{i}\frac{\delta_\ell}{\delta X_a'(t)}\ \frac{1}{i}\frac{\delta_\ell}{\delta X_b'(t')} \langle t_1|t_2\rangle = \varepsilon_{ab}(t\ ,\ t')\ \langle t_1|\ \left(x_a(t)x_b(t')\right)_+\ |t_2\rangle \tag{7.74}$$

where

$$\varepsilon_{k\ell} = \varepsilon_{k\alpha} = \varepsilon_{\alpha k} = 1$$

$$\varepsilon_{\alpha\beta}(t\ ,\ t') = +1\ ,\quad t > t' \qquad (7.75)$$

$$= -1\ ,\quad t < t'\ .$$

It follows from the properties of the $\delta X'$, or explicitly from (7.74), that variational derivatives are commutative, with the exception of those referring entirely to variables of the second kind, which are anticommutative.

In the limit as $t \to t'$ in a definite sense, we obtain operator products referring to a common time in which the order of multiplication is still determined by the time order. Thus

$$\langle t_1|x_a(t)x_b(t)|t_2\rangle = \frac{1}{i}\frac{\delta_\ell}{\delta X_a'(t)}\frac{1}{i}\frac{\delta_\ell}{\delta X_b'(t-0)}\langle t_1|t_2\rangle\ , \qquad (7.76)$$

whereas

$$\pm\ \langle t_1|x_b(t)x_a(t)|t_2\rangle = \frac{1}{i}\frac{\delta_\ell}{\delta X_a'(t)}\frac{1}{i}\frac{\delta_\ell}{\delta X_b'(t+0)}\langle t_1|t_2\rangle\ , \qquad (7.77)$$

in which the minus sign appears only for pairs of second kind variables. The difference between the two limits is related to the commutation properties of the fundamental dynamical variables

$$\langle t_1 | \{[x_a(t) , x_b(t)]\} | t_2 \rangle$$

$$= \frac{\delta_\ell}{\delta X_a'(t)} \left(\frac{\delta_\ell}{\delta X_b'(t+0)} - \frac{\delta_\ell}{\delta X_b'(t-0)} \right) \langle t_1 | t_2 \rangle \tag{7.78}$$

$$= i \frac{\delta_\ell}{\delta X_a'(t)} \langle t_1 | \rho\big(x_b(t+0) - x_b(t-0)\big) | t_2 \rangle$$

where the ambiguous bracket implies the commutator, for variables of the first kind or a single variable of the second kind, and becomes the anticommutator for variables of the second kind. Now according to the equations of motion for the system with Hamiltonian $H - \sum X_a x_a$,

$$- \frac{dx}{dt} A = \frac{\partial_r H}{\partial x} - X \quad , \tag{7.79}$$

the change of the dynamical variables in a small time interval, as determined by the displacement, is

$$x(t+0) - x(t-0) = \int_{t-0}^{t+0} dt' \; X(t') \; A^{-1} \quad . \tag{7.80}$$

Hence

$$\frac{\delta_\ell}{\delta X'_a(t)} \langle t_1|\rho\{x_b(t+0) - x_b(t-0)\}|t_2\rangle = (A^{-1})_{ab}\langle t_1|t_2\rangle \, , \tag{7.81}$$

and

$$\{[x_a(t) \, , \, x_b(t)]\} = i \, (A^{-1})_{ab} \tag{7.82}$$

which combines the commutation properties of all the fundamental variables.

Products of three operators at a common time are expressed, by a formula of the type (7.76), as

$$\langle t_1|\rho x_a(t) x_b(t) x_c(t)|t_2\rangle$$
$$= \frac{1}{i}\frac{\delta_\ell}{\delta X'_a(t+0)} \frac{1}{i}\frac{\delta_\ell}{\delta X'_b(t)} \frac{1}{i}\frac{\delta_\ell}{\delta X'_c(t-0)} \langle t_1|t_2\rangle \tag{7.83}$$

and more generally, if $F\{x(t)\}$ is any algebraic function of the dynamical variables at time t, restricted only to be an even function of variables of the second kind, we have

$$\langle t_1|F\{x(t)\}|t_2\rangle = F\left(\frac{1}{i}\frac{\delta_\ell}{\delta X'(t)} \right) \langle t_1|t_2\rangle \, , \tag{7.84}$$

in which the variational derivatives refer to times differing infinitesimally from t as implied by the particular multiplication order of the operators. If the Hamiltonian operator is an algebraic function of the dynamical variables, we can utilize this differential operator representation through the device of considering a related system with the Hamiltonian λH, and examining the effect of an infinitesimal change in the parameter λ,

$$\begin{aligned} i \frac{\partial}{\partial \lambda} \langle t_1 | t_2 \rangle_X &= \left\langle t_1 \right| \int_{t_2}^{t_1} dt \, H\big(x(t)\big) \left| t_2 \right\rangle_X \\ &= \int_{t_2}^{t_1} dt \, H\left(\frac{1}{i} \frac{\delta_\ell}{\delta X'(t)} \right) \langle t_1 | t_2 \rangle_X \quad . \end{aligned} \tag{7.85}$$

This differential equation is analogous in structure to a Schrödinger equation, with the transformation function, in its dependence upon the infinite number of variables $X_a'(t)$, $t_1 \geq t \geq t_2$, appearing as the wave function. On integrating from $\lambda = 0$ to $\lambda=1$, the formal solution,

$$\langle t_1|t_2\rangle_{HX} = \exp\left[-i\int_{t_2}^{t_1} dt\, H\left(\frac{1}{i}\frac{\delta_\ell}{\delta X'(t)}\right)\right] \langle t_1|t_2\rangle_{OX} \tag{7.86}$$

exhibits the transformation function as obtained by a process of differentiation from the elementary transformation function that refers only to the special canonical transformations. A more general form emerges from the decomposition

$$H = H^0 + H^1 \quad . \tag{7.87}$$

It must be noted that, since H^0 and H^1 are necessarily even functions of the variables of the second kind, the associated differential operators are commutative. On applying first the exponential operator constructed from H^0, we obtain

$$\langle t_1|t_2\rangle_{HX} \tag{7.88}$$

$$= \exp\left[-i\int_{t_2}^{t_1} dt\, H^1\left(\frac{1}{i}\frac{\delta_\ell}{\delta X'(t)}\right)\right] \left\langle t_1\middle|t_2\right\rangle_{H^0X}$$

which could also be derived directly by appropriate modification of (7.52). This produces a basis for perturbation theory, by which the desired transformation function for Hamiltonian H is obtained from a simpler one for Hamiltonian H^0 by repeated differentiation.

7.9 INTERACTION OF TWO SUBSYSTEMS

The general dynamical situation of a system formed of two sub-systems in interaction is described by

$$H^0 = H_1(x_1) + H_2(x_2) \quad , \qquad H^1 = H_{12}(x_1 , x_2) \tag{7.89}$$

where x_1 and x_2 refer to the dynamical variables of the respective sub-systems. Correspondingly the displacements X decompose into two sets X_1 and X_2. In the non-interacting system described by the Hamiltonian H^0, the two sub-systems are dynamically independent and, in accordance with the additive structure of the action operator, the transformation function appears as the product of that for the separate sub-systems,

$$\langle t_1 | t_2 \rangle_{H^0 X} = \langle t_1 | t_2 \rangle_{H_1 X_1} \langle t_1 | t_2 \rangle_{H_2 X_2} \quad . \tag{7.90}$$

The transformation function for the interacting systems is therefore given by

$$\langle t_1 | t_2 \rangle_{HX}$$

$$= \exp\left[-i \int_{t_2}^{t_1} dt \, H_{12} \left(\frac{1}{i} \frac{\delta_\ell}{\delta X_1'} , \frac{1}{i} \frac{\delta_\ell}{\delta X_2'} \right) \right] \tag{7.91}$$

$$\times \left\langle t_1 \middle| t_2 \right\rangle_{H_1 X_1} \left\langle t_1 \middle| t_2 \right\rangle_{H_2 X_2} \quad .$$

In this construction, both sub-systems appear quite symmetrically. It is often convenient, however, to introduce an asymmetry in viewpoint whereby one part of the system is thought of as moving under the influence of the other. The Hamiltonian $H_1(x_1) + H_{12}(x_1 , \xi_2(t))$ describes the first system only, as influenced by the external disturbance originating in the second system, with its variables regarded as prescribed but arbitrary functions of the time. The transformation function for this incomplete system will be given by

$$\left\langle t_1 \middle| t_2 \right\rangle_{H_1+H_{12}(\xi_2),\, X_1}$$

$$= \exp\left[-i \int_{t_2}^{t_1} dt\, H_{12}\left(\frac{1}{i} \frac{\delta_\ell}{\delta X_1'(t)} ,\ \xi_2(t)\right)\right] \tag{7.92}$$

$$\times \left\langle t_1 \middle| t_2 \right\rangle_{H_1 X_1}$$

and (7.91) asserts that the complete transformation function can be obtained by replacing the prescribed variables of the second system by differential operators that act on the transformation function referring to the second system without interaction,

$$\left\langle t_1 \middle| t_2 \right\rangle_{HX} \tag{7.93}$$

$$= \left\langle t_1 \middle| t_2 \right\rangle_{H_1 + H_{12}\left(-i \frac{\delta_\ell}{\delta X_2'}\right),\, X_1} \left\langle t_1 \middle| t_2 \right\rangle_{H_2 X_2} .$$

The displacements have served their purpose, after the differentiations are performed, and will be placed equal to zero throughout the interval between t_1 and t_2 . They will still be needed at the terminal times if the transformation functions refer to standard eigenvalues, but will be set equal to zero everywhere if this device is not employed. Considering the latter for simplicity of notation, the transformation function of the system with Hamiltonian H is obtained as

$$\langle t_1 | t_2 \rangle_H \tag{7.94}$$

$$= \langle t_1 | t_2 \rangle_{H_1 + H_{12}\left(-i \frac{\delta_\ell}{\delta X_2'}\right)} \langle t_1 | t_2 \rangle_{H_2 X_2} \Bigg|_{X_2'=0}$$

which has the form of a scalar product evaluated by the differential composition of wave functions. Accordingly, from the known equivalent evaluations of such products, other forms can be given to (7.94), such as

$$\langle t_1 | t_2 \rangle_H \tag{7.95}$$

$$= \langle t_1 | t_2 \rangle_{H_1 + H_{12}(\xi_2)} \langle t_1 | t_2 \rangle_{H_2 ,\ -i \frac{\delta_r^T}{\delta \xi_2}} \Bigg|_{\xi_2=0}$$

and, for variables of the first kind, one can substitute composition by integration for the differential method here employed.

7.10 ADDENDUM: EXTERIOR ALGEBRA AND THE ACTION PRINCIPLE†

---

† Reproduced from the Proceedings of the National Academy of Sciences, Vol. 48 pp. 603-611 (1962).

Reprinted from the Proceedings of the NATIONAL ACADEMY OF SCIENCES
Vol. 48, No. 4, pp. 603–611. April, 1962.

*EXTERIOR ALGEBRA AND THE ACTION PRINCIPLE, I**

BY JULIAN SCHWINGER

HARVARD UNIVERSITY

Communicated February 27, 1962

The quantum action principle[1] that has been devised for quantum variables of type $\nu = 2$ lacks one decisive feature that would enable it to function as an instrument of calculation. To overcome this difficulty we shall enlarge the number system by adjoining an exterior or Grassmann algebra.[2]

An exterior algebra is generated by N elements ϵ_κ, $\kappa = 1 \cdot\cdot N$, that obey

$$\{\epsilon_\kappa, \epsilon_\lambda\} = 0,$$

which includes $\epsilon_\kappa^2 = 0$.

A basis for the exterior algebra is supplied by the unit element and the homogeneous products of degree d

$$\epsilon_{\kappa_1}\epsilon_{\kappa_2}\cdot\cdot\epsilon_{\kappa_d}, \quad \kappa_1 < \kappa_2 \cdot\cdot < \kappa_d$$

for $d = 1 \cdot\cdot N$. The total number of linearly independent elements is counted as

$$\sum_{d=0}^{N} \frac{N!}{d!(N-d)!} = 2^N.$$

The algebraic properties of the generators are unaltered by arbitrary nonsingular linear transformations.

To suggest what can be achieved in this way, we consider a new class of special variations, constructed as the product of ξ_{2n+1} with an arbitrary real infinitesimal linear combination of the exterior algebra generators,

$$^1/_2\delta\xi_k = -\xi_{2n+1}\sum_{\kappa=1}^{N}\delta\omega_{k\kappa}\epsilon_\kappa.$$

The members of this class anticommute with the ξ_k variables, owing to the factor ξ_{2n+1}, and among themselves,

$$\{\delta^{(1)}\xi_k,\quad \delta^{(2)}\xi_l\} = 0,$$

since they are linear combinations of the exterior algebra generators. Accordingly, the generator of a special variation,

$$G_\xi = {}^1/_2 i\sum_k \xi_k\delta\xi_k = -{}^1/_2 i\sum \delta\xi_k\xi_k$$
$$= \sum_{k\kappa}\delta\omega_{k\kappa}\epsilon_\kappa(1/i)\xi_k\xi_{2n+1},$$

commutes with any such variation,

$$[\delta^{(1)}\xi_k,\ G_\xi^{(2)}] = 0.$$

If we consider the commutator of two generators we get

$$(1/i)[G_\xi^{(1)},\ G_\xi^{(2)}] = -{}^1/_2 i\sum\delta^{(1)}\xi_k(1/i)[\xi_k,\ G_\xi^{(2)}] = -{}^1/_4 i\sum\delta^{(1)}\xi_k\delta^{(2)}\xi_k,$$

where the right-hand side is proportional to the unit operator, and to a bilinear function of the exterior algebra generating elements,

$$\sum\delta^{(1)}\xi_k\delta^{(2)}\xi_k = \sum(\delta^{(1)}\omega_{k\lambda}\delta^{(2)}\omega_{k\mu} - \delta^{(1)}\omega_{k\mu}\delta^{(2)}\omega_{k\lambda})\epsilon_\lambda\epsilon_\mu.$$

The latter structure commutes with all operators and with all exterior algebra elements. The commutator therefore commutes with any generator G_ξ, and the totality of the new special variations has a group structure which is isomorphic to that of the special canonical group for the degrees of freedom of type $\nu = \infty$. This, rather than the rotation groups previously discussed, is the special canonical group for the $\nu = 2$ variables.

We are thus led to reconsider the action principle, now using the infinitesimal variations of the special canonical group. The class of transformation generators G that obey

$$[\delta\xi_k,\ G] = 0$$

includes not only all even operator functions of the $2n$ variables ξ_k, but also even functions of the ξ_k that are multiplied by even functions of the N exterior algebra generators ϵ_κ, and odd functions of the ξ_k multiplied by odd functions of the ϵ_κ. The generators of the special canonical transformations are included in the last category.

The concept of a Hermitian operator requires generalization to accommodate the noncommutative numerical elements. The reversal in sense of multiplication that is associated with the adjoint operation now implies that

$$(\lambda A)^\dagger = A^\dagger\lambda^*,$$

where λ is a number, and we have continued to use the same notation and language despite the enlargement of the number system. Complex conjugation thus has the algebraic property

$$(\lambda_1\lambda_2)^* = \lambda_2^*\lambda_1^*.$$

The anticommutativity of the generating elements is maintained by this operation, and we therefore regard complex conjugation in the exterior algebra as a linear mapping of the N-dimensional subspace of generators,

$$\epsilon_\kappa^* = \sum_\lambda R_{\kappa\lambda}\epsilon_\lambda.$$

The matrix R obeys

$$R^*R = 1,$$

which is not a statement of unitarity. Nevertheless, a Cayley parametrization exists,

$$R = \frac{1 + ir}{1 - ir}, \quad r^* = r,$$

provided $\det(1 + R) \neq 0$. Then,

$$\bar{\epsilon}_\kappa = \epsilon_\kappa + i \sum r_{\kappa\lambda}\epsilon_\lambda$$

obeys

$$\bar{\epsilon}_\kappa^* = \bar{\epsilon}_\kappa,$$

which asserts that a basis can always be chosen with real generators. There still remains the freedom of real nonsingular linear transformations.

This conclusion is not altered when R has the eigenvalue -1. In that circumstance, we can construct $p(R)$, a polynomial in R that has the properties

$$(1 + R)p = 0, \quad p(1 - p) = 0, \quad p^* = p, \quad \text{and} \quad p(-1) = 1.$$

The matrix

$$R' = R(1 - 2p) = R(1 - p) + p$$

also obeys

$$R'^*R' = 1,$$

while

$$\det(1 + R') \neq 0,$$

since the contrary would imply the existence of a nontrivial vector v such that

$$(1 + R')v = ((1 + R)(1 - p) + 2p)v = 0$$

or, equivalently,

$$pv = 0, \quad (1 + R)v = 0,$$

which is impossible since $p(-1) = 1$. Now,

$$1 - 2p = \frac{1 - p - ip}{1 - p + ip},$$

and if we use the Cayley construction for R' in terms of a real matrix r', which is a function of R with the property $r'(-1) = 0$, we get

$$R = \frac{1 - p + ip}{1 - p - ip}\frac{1 + ir'}{1 - ir'} = \frac{1 - p + i(p + (1 - p)r')}{1 - p - i(p + (1 - p)r')}.$$

This establishes the generality of the representation

$$R = \rho/\rho^*,$$

where the nonsingular matrices ρ and ρ^* are commutative, and thereby proves the reality of the generator set

$$\bar{\epsilon}_\kappa = \sum \rho_{\kappa\lambda}\epsilon_\lambda.$$

When real generators are chosen, the other elements of a real basis are supplied by the nonvanishing products

$$i^{d(d-1)/2}\,\epsilon_{\kappa 1}\cdots\epsilon_{\kappa d},$$

for

$$(\epsilon_{\kappa 1}\cdots\epsilon_{\kappa d})^* = \epsilon_{\kappa d}\cdots\epsilon_{\kappa 1}$$
$$= (-1)^{d(d-1)/2}\,\epsilon_{\kappa 1}\cdots\epsilon_{\kappa d}.$$

A Hermitian operator, in the extended sense, is produced by linear combinations of conventional Hermitian operators multiplied by real elements of the exterior algebra. The generators G_ξ are Hermitian, as are the commutators $i[G_\xi^{(1)}, G_\xi^{(2)}]$.

The transformation function $\langle\tau_1|\tau_2\rangle$ associated with a generalized unitary transformation is an element of the exterior algebra. It possesses the properties

$$\langle\tau_1|\tau_2\rangle^* = \langle\tau_2|\tau_1\rangle$$

and

$$\langle\tau_1|\tau_3\rangle = \langle\tau_1|\tau_2\rangle \times \langle\tau_2|\tau_3\rangle,$$

where $\times$ symbolizes the summation over a complete set of states which, for the moment at least, are to be understood in the conventional sense. These attributes are consistent with the nature of complex conjugation, since

$$(\langle\tau_1|\tau_2\rangle \times \langle\tau_2|\tau_3\rangle)^* = \langle\tau_3|\tau_2\rangle \times \langle\tau_2|\tau_1\rangle = \langle\tau_3|\tau_1\rangle.$$

For an infinitesimal transformation, we have

$$\langle\tau + d\tau|\tau\rangle = \langle|1 + id\tau G(\xi, \tau)|\rangle,$$

where G is Hermitian, which implies that its matrix array of exterior algebra elements obeys

$$\langle a'|G|a''\rangle^* = \langle a''|G|a'\rangle.$$

The subsequent discussion of the action principle requires no explicit reference to the structure of the special variations, and the action principle thus acquires a dual significance, depending upon the nature of the number system.

We shall need some properties of differentiation in an exterior algebra. Since $\epsilon_\lambda^2 = 0$, any given function of the generators can be displayed uniquely in the alternative forms

$$f(\epsilon) = f_0 + \epsilon_\lambda f_l = f_0 + f_r\epsilon_\lambda,$$

where f_0, f_l, f_r do not contain ϵ_λ. By definition, f_r and f_l are the right and left derivatives, respectively, of $f(\epsilon)$ with respect to ϵ_λ,

$$\frac{\partial_l f(\epsilon)}{\partial \epsilon_\lambda} = f_l, \quad \frac{\partial_r f(\epsilon)}{\partial \epsilon_\lambda} = f_r.$$

If $f(\epsilon)$ is homogeneous of degree d, the derivatives are homogeneous of degree $d - 1$, and left and right derivatives are equal for odd d, but of opposite sign when d is even. For the particular odd function ϵ_μ, we have

$$\epsilon_\mu = \epsilon_\mu(1 - \delta_{\lambda\mu}) + \epsilon_\lambda \delta_{\lambda\mu},$$

so that

$$\partial \epsilon_\mu / \partial \epsilon_\lambda = \delta_{\lambda\mu}.$$

Since a derivative is independent of the element involved, repetition of the operation annihilates any function,

$$(\partial/\partial\epsilon_\lambda)^2 f(\epsilon) = 0.$$

To define more general second derivatives we write, for $\lambda \neq \mu$,

$$f(\epsilon) = f_0 + \epsilon_\lambda f_\lambda + \epsilon_\mu f_\mu + \epsilon_\lambda \epsilon_\mu f_{\lambda\mu},$$

where the coefficients are independent of ϵ_λ and ϵ_μ. The last term has the alternative forms

$$\epsilon_\lambda \epsilon_\mu f_{\lambda\mu} = \epsilon_\mu \epsilon_\lambda f_{\mu\lambda}$$

with

$$f_{\mu\lambda} = -f_{\lambda\mu}.$$

Then,

$$\frac{\partial_l f}{\partial \epsilon_\lambda} = f_\lambda + \epsilon_\mu f_{\lambda\mu}, \quad \frac{\partial_l f}{\partial \epsilon_\mu} = f_\mu + \epsilon_\lambda f_{\mu\lambda}$$

and

$$\frac{\partial_l}{\partial \epsilon_\mu} \frac{\partial_l}{\partial \epsilon_\lambda} f = f_{\lambda\mu}, \quad \frac{\partial_l}{\partial \epsilon_\lambda} \frac{\partial_l}{\partial \epsilon_\mu} f = f_{\mu\lambda},$$

which shows that different derivatives are anticommutative,

$$\left\{ \frac{\partial_l}{\partial \epsilon_\mu}, \frac{\partial_l}{\partial \epsilon_\lambda} \right\} f(\epsilon) = 0.$$

A similar statement applies to right derivatives.

The definition of the derivative has been given in purely algebraic terms. We now consider $f(\epsilon + \delta\epsilon)$, where $\delta\epsilon_\lambda$ signifies a linear combination of the exterior algebra elements with arbitrary infinitesimal numerical coefficients, and conclude that

$$f(\epsilon + \delta\epsilon) - f(\epsilon) = \sum \delta\epsilon_\lambda \frac{\partial_l f}{\partial \epsilon_\lambda} = \sum \frac{\partial_r f}{\partial \epsilon_\lambda} \delta\epsilon_\lambda,$$

to the first order in the infinitesimal numerical parameters. If this differential property is used to identify derivatives, it must be supplemented by the requirement that the derivative be of lower degree than the function, for any numerical multiple of $\epsilon_1 \epsilon_2 \cdots \epsilon_N$ can be added to a derivative without changing the differential form. Let us also note the possibility of using arbitrary nonsingular linear combinations of the ϵ_λ in differentiation, as expressed by the matrix formula

$$(\partial/\partial a\epsilon)f = (a^{-1})^T(\partial/\partial\epsilon)f,$$

which follows directly from the differential expression.

We shall now apply the extended action principle to the superposition of two transformations, one produced by a conventional Hermitian operator G, an even function of the ξ_k, while the other is a special canonical transformation performed arbitrarily, but continuously in τ. The effective generator is

$$G(\xi(\tau), \tau) - i\sum_{k=1}^{2n} \xi_k(\tau)X_k(\tau),$$

where $-X_k(\tau)d\tau$ is the special variation induced in $\xi_k(\tau)$ during the interval $d\tau$. The objects $X_k(\tau)$ are constructed by multiplying ξ_{2n+1} with a linear combination of the ϵ_λ containing real numerical coefficients that are arbitrary continuous functions of τ. To use the stationary action principle, we observe that each $X_k(\tau)$, as a special variation, is commutative with a generator of special variations and thus

$$\delta' X_k(\tau) = 0.$$

The action principle then asserts that

$$\sum_k \delta\xi_k\left[i\,\frac{d\xi_k}{d\tau} + \frac{\partial_l G}{\partial \xi_k} - iX_k\right] = 0,$$

where the $\delta\xi_k$ are variations constructed from the exterior algebra elements. We cannot entirely conclude that

$$\frac{d\xi_k}{d\tau} = i\,\frac{\partial_l G}{\partial \xi_k} + X_k,$$

since there remains an arbitrariness associated with multiples of $\epsilon_1 \cdots \epsilon_N$, as in the identification of derivatives from a differential form. No such term appears, however, on evaluating

$$\frac{d\xi_k}{d\tau} = -\frac{1}{i}\left[\xi_k, G - i\sum \xi_l X_l\right] = i\,\frac{\partial_l G}{\partial \xi_k} + X_k.$$

This apparent incompleteness of the action principle is removed on stating the obvious requirement that the transformation function $\langle \tau_1 | \tau_2 \rangle$ be an ordinary number when all $X_k(\tau)$ vanish. Thus, the elements of the exterior algebra enter only through the products of the $X_k(\tau)$ with $\xi_k(\tau)$, as the latter are obtained by integrating the equations of motion, and terms in these equations containing $\epsilon_1 \cdots \epsilon_N$ are completely without effect.

Let us examine how the transformation function $\langle \tau_1 | \tau_2 \rangle_G{}^X$ depends upon the $X_k(\tau)$. It is well to keep in mind the two distinct factors that compose $X_k(\tau)$,

$$X_k(\tau) = \rho X_k'(\tau).$$

Here,
$$\rho = 2^{1/2}\xi_{2n+1}, \quad \rho^2 = 1,$$

while $X_k'(\tau)$ is entirely an element of the exterior algebra. Thus, it is really the $X'(\tau)$ upon which the transformation function depends. An infinitesimal change of the latter induces

$$\delta\langle\tau_1|\tau_2\rangle_G{}^X = \langle\tau_1|\int_{\tau_2}^{\tau_1} d\tau\xi(\tau)\delta X(\tau)|\tau_2\rangle_G{}^X = \int_{\tau_2}^{\tau_1} d\tau\langle\tau_1|\xi\rho\delta X'|\tau_2\rangle_G{}^X,$$

where the summation index k has been suppressed. The repetition of such variations gives the ordered products

$$\delta^{2m}\langle\tau_1|\tau_2\rangle = \int_{\tau_2}^{\tau_1} d\tau^1\cdot\cdot d\tau^{2m}\langle\tau_1|(\xi\delta X(\tau^1)\cdot\cdot\xi\delta X(\tau^{2m}))_0|\tau_2\rangle = \int d\tau^1\cdot\cdot d\tau^{2m}(\delta X'(\tau^1)\cdot\cdot\delta X'(\tau^{2m}))_0{}^*\langle\tau_1|(\xi(\tau^1)\cdot\cdot\xi(\tau^{2m}))_0|\tau_2\rangle,$$

which form is specific to an even number of variations. Complex conjugation is applied to the exterior algebra elements in order to reverse the sense of multiplication. In arriving at the latter form, we have exploited the fact that special canonical variations are not implicit functions of τ and therefore anticommute with the ξ_k without regard to the τ values. Thus, one can bring together the $2m$ quantities $\delta X(\tau^1)$, $\cdot\cdot\delta X(\tau^{2m})$, and this product is a multiple of the unit operator since $\rho^2 = 1$. The multiple is the corresponding product of the exterior algebra elements $\delta X'(\tau^1)$, $\cdot\cdot\delta X'(\tau^{2m})$, which, as an even function, is completely commutative with all elements of the exterior algebra and therefore can be withdrawn from the matrix element. The reversal in multiplication sense of the exterior algebra elements gives a complete account of the sign factors associated with anticommutativity.

The notation of functional differentiation can be used to express the result. With left derivatives, we have

$$\left(\frac{\delta_l}{\delta X'(\tau^1)}\cdot\cdot\frac{\delta_l}{\delta X'(\tau^{2m})}\right)_0\langle\tau_1|\tau_2\rangle^X = \langle\tau_1|(\xi(\tau^1)\cdot\cdot\xi(\tau^{2m}))_0|\tau_2\rangle^X,$$

where the anticommutativity of exterior algebra derivatives implies that

$$\left(\frac{\delta}{\delta X'(\tau^1)}\cdot\cdot\frac{\delta}{\delta X'(\tau^{2m})}\right)_0 = \epsilon_0(\tau^1\cdot\cdot\tau^{2m})\frac{\delta}{\delta X'(\tau^1)}\cdot\cdot\frac{\delta}{\delta X'(\tau^{2m})}.$$

Here,

$$\epsilon_0(\tau^1\cdot\cdot\tau^{2m}) = \pm 1$$

according to whether an even or odd permutation is required to bring τ^1, $\cdot\cdot\tau^{2m}$ into the ordered sequence. This notation ignores one vital point, however. In an exterior algebra with N generating elements, no derivative higher than the Nth exists. If we wish to evaluate unlimited numbers of derivatives, in order to construct correspondingly general functions of the dynamical variables, we must choose $N = \infty$; the exterior algebra is of infinite dimensionality. Then we can assert of arbitrary even ordered functions of the $\xi_k(\tau)$ that

$$\langle\tau_1|(F(\xi))_0|\tau_2\rangle^X = F(\delta_l/\delta X')_0\langle\tau_1|\tau_2\rangle^X = \langle\tau_1|\tau_2\rangle^X F(\delta_r{}^T/\delta X')_0.$$

In the alternative right derivative form, δ^T signifies that successive differentiations are performed from left to right rather than in the conventional sense. We also note that the particular order of multiplication for operator products at a common time is to be produced by limiting processes from unequal τ values. As an application to the transformation function $\langle\tau_1|\tau_2\rangle_G{}^X$, we supply the even operator function G with a variable factor λ and compute

$$\frac{\partial}{\partial\lambda}\langle\tau_1|\tau_2\rangle_{\lambda G}^X = i\langle\tau_1|\int_{\tau_2}^{\tau_1} d\tau G(\xi, \tau)|\tau_2\rangle = i\int_{\tau_2}^{\tau_1} d\tau\, G\left(\frac{\delta_l}{\delta X'}, \tau\right)\langle\tau_1|\tau_2\rangle,$$

which gives the formal construction

$$\langle\tau_1|\tau_2\rangle_G^X = \exp[i\int_{\tau_2}^{\tau_1} d\tau G(\delta_l/\delta X', \tau)]\langle\tau_1|\tau_2\rangle^X,$$

where the latter transformation function refers entirely to the special canonical group.

The corresponding theorems for odd ordered functions of the $\xi_k(\tau)$ are

$$F(\delta_l/\delta X')_0\langle\tau_1|\tau_2\rangle^X = -\langle\tau_1|\rho(\tau_1)(F(\xi))_0|\tau_2\rangle^X$$

and

$$\langle\tau_1|\tau_2\rangle^X F(\delta_r{}^T/\delta X')_0 = \langle\tau_1|(F(\xi))_0\rho(\tau_2)|\tau_2\rangle^X,$$

provided the states are conventional ones. Since the variations $\delta X(\tau)$ have no implicit τ dependence, the factors $\rho\epsilon$ can be referred to any τ value. If this is chosen as τ_1 or τ_2 there will remain one ρ operator while the odd product of exterior algebra elements can be withdrawn from the matrix element if, as we assume, the product commutes with the states $\langle\tau_1|$ and $|\tau_2\rangle$. The sign difference between the left and right derivative forms stems directly from the property $\xi\delta X = -\delta X\xi$.

As an example, let $F(\xi)$ be the odd function that appears on the left-hand side in the equation of motion

$$\frac{d\xi}{d\tau} - i\frac{\partial_l G}{\partial\xi} = X.$$

The corresponding functional derivatives are evaluated as

$$\langle\tau_1|\rho(\tau_1)X(\tau)|\tau_2\rangle = X'(\tau)\langle\tau_1|\tau_2\rangle$$

or

$$\langle\tau_1|X(\tau)\rho(\tau_2)|\tau_2\rangle = \langle\tau_1|\tau_2\rangle X'(\tau),$$

since the $X(\tau)$ are special canonical displacements, and this yields the functional differential equations obeyed by a transformation function $\langle\tau_1|\tau_2\rangle_G{}^X$, namely,

$$\left[-\frac{d}{d\tau}\frac{\delta_l}{\delta X'(\tau)} + i\frac{\partial_l G}{\partial\xi}\left(\frac{\delta_l}{\delta X'}\right) - X'(\tau)\right]\langle\tau_1|\tau_2\rangle_G{}^X = 0$$

and

$$\langle\tau_1|\tau_2\rangle_G{}^X\left[\frac{d}{d\tau}\frac{\delta_r{}^T}{\delta X'(\tau)} + i\frac{\partial_r G}{\partial\xi}\left(\frac{\delta_r{}^T}{\delta X'}\right) - X'(\tau)\right] = 0.$$

Some properties that distinguish the trace of the transformation function can also be derived from the statement about odd functions. Thus,

$$tr\langle\tau_1|\rho(\tau_1)F|\tau_2\rangle = tr\langle\tau_1|F\rho(\tau_2)|\tau_2\rangle,$$

since either side is evaluated as

$$\sum_{a'a''}\langle a'|\rho|a''\rangle\langle a''\tau_1|F|a'\tau_2\rangle.$$

Accordingly,

$$F(\delta_l/\delta X')_0\, tr\,\langle\tau_1|\,\tau_2\rangle^X = -tr\langle\tau_1|\,\tau_2\rangle^X F(\delta_r{}^T/\delta X')_0$$

and, in particular,

$$\left(\frac{\delta_l}{\delta X'(\tau)} + \frac{\delta_r}{\delta X'(\tau)}\right) tr\langle\tau_1|\,\tau_2\rangle_G{}^X = 0,$$

which shows that the trace is an even function of the $X'(\tau)$. The nature of the trace is involved again in the statement

$$tr\langle\tau_1|\,\rho(\tau_1)(\xi(\tau_1) + \xi(\tau_2))\,|\,\tau_2\rangle^X = tr\langle\tau_1|\,\{\,\rho(\tau_1),\ \xi(\tau_1)\}\,|\,\tau_2\rangle = 0,$$

which is an assertion of effective antiperiodicity for the operators $\xi(\tau)$ over the interval $T = \tau_1 - \tau_2$. The equivalent restriction on the trace of the transformation function is

$$\left(\frac{\delta_l}{\delta X'(\tau_1)} + \frac{\delta_l}{\delta X'(\tau_2)}\right) tr\,\langle\tau_1|\,\tau_2\rangle_G{}^X = 0,$$

or

$$\left(\frac{\delta_l}{\delta X'(\tau_1)} - \frac{\delta_r}{\delta X'(\tau_2)}\right) tr\,\langle\tau_1|\,\tau_2\rangle_G{}^X = 0.$$

* Publication assisted by the Air Force Office of Scientific Research.

[1] These Proceedings, **47**, 1075 (1961).

[2] A brief mathematical description can be found in the publication *The Construction and Study of Certain Important Algebras*, by C. Chevalley (1955, the Mathematical Society of Japan). Although such an extension of the number system has long been employed in quantum field theory (see, for example, these Proceedings, **37**, 452 (1951)), there has been an obvious need for an exposition of the general algebraic and group theoretical basis of the device.

CHAPTER EIGHT
GREEN'S FUNCTIONS

8.1 INCORPORATION OF INITIAL CONDITIONS

The most elementary method for the construction of canonical transformation functions associated with parameterized transformations is the direct solution of the differential equations that govern the dependence upon the parameters. For time development these are the Schrödinger equations

$$\left(i \frac{\partial}{\partial t_1} - H\right) \langle t_1 | t_2 \rangle = 0 \quad , \quad \langle t_1 | t_2 \rangle \left(-i \frac{\partial^T}{\partial t_2} - H\right) = 0 \tag{8.1}$$

where H here refers to the differential operator representatives of the Hamiltonian, at times t_1 or t_2, which depend upon the particular choice of canonical representation. The desired transformation function is distinguished among the solutions of these Schrödinger equations by the initial condition referring to equal times,

$$\langle t | t \rangle = \langle | \rangle \quad , \tag{8.2}$$

which means that the canonical transformation function is independent of the common time and is determined only by the relation between the descriptions. This formalism is given an operator basis on writing

$$\langle t_1 | t_2 \rangle = \langle | U(t_1 , t_2) | \rangle \tag{8.3}$$

where the unitary time development operator

$$U(t_1 , t_2)^{\dagger} = U(t_2 , t_1) = U(t_1 , t_2)^{-1} \tag{8.4}$$

is to be constructed as a function of dynamical variables that do not depend upon time, by solving the differential equations

$$\left(i \frac{\partial}{\partial t_1} - H\right) U(t_1 , t_2) = 0 \quad ,$$
$$U(t_1 , t_2)\left(-i \frac{\partial^T}{\partial t_2} - H\right) = 0 \qquad (8.5)$$

together with the initial condition

$$U(t , t) = 1 \quad . \qquad (8.6)$$

It is useful to incorporate the initial conditions that characterize transformation functions, or the time development operator, into the differential equations. This is accomplished by introducing related discontinuous functions of time - the Green's functions or operators. The retarded and advanced Green's functions are the matrices, in some representation, of the retarded and advanced Green's operators defined, respectively, by

$$G_r(t_1 , t_2) = -i\ \eta_+(t_1 , t_2)\ U(t_1 , t_2)$$
$$G_a(t_1 , t_2) = i\ \eta_-(t_1 , t_2)\ U(t_1 , t_2) \qquad (8.7)$$

where

$$\eta_+(t_1\ ,\ t_2) = 1\ ,\quad t_1 > t_2$$
$$= 0\ ,\quad t_1 < t_2 \qquad (8.8)$$

and

$$\eta_-(t_1\ ,\ t_2) = \eta_+(t_2\ ,\ t_1) = 1 - \eta_+(t_1\ ,\ t_2)\ . \qquad (8.9)$$

The discontinuities of the functions η_+ and η_- are expressed in differential form by

$$\frac{\partial}{\partial t_1}\eta_\pm(t_1\ ,\ t_2) = -\ \eta_\pm(t_1\ ,\ t_2)\frac{\partial^T}{\partial t_2} = \pm\ \delta(t_1 - t_2) \qquad (8.10)$$

and therefore, in consequence of the differential equations (8.5) and the initial conditions (8.6), both the retarded and advanced Green's operator obey the inhomogeneous equations

$$\left(i\frac{\partial}{\partial t_1} - H\right)G(t_1\ ,\ t_2)$$
$$= G(t_1\ ,\ t_2)\left(-i\frac{\partial^T}{\partial t_2} - H\right) = \delta(t_1 - t_2)\ . \qquad (8.11)$$

The two Green's operators are distinguished as the solutions of these equations that obey

$$G_r(t_1\,,\,t_2) = 0 \quad , \qquad t_1 < t_2$$
$$G_a(t_1\,,\,t_2) = 0 \quad , \qquad t_1 > t_2 \tag{8.12}$$

which is evidently consistent with the adjoint relation

$$G_r(t_1\,,\,t_2)^{\dagger} = G_a(t_2\,,\,t_1) \quad . \tag{8.13}$$

From these operators the unitary time development operator is constructed as

$$U(t_1\,,\,t_2) = i[G_r(t_1\,,\,t_2) - G_a(t_1\,,\,t_2)] \quad . \tag{8.14}$$

8.2 CONSERVATIVE SYSTEMS. TRANSFORMS

For a conservative system, in which t does not appear in the Hamiltonian operator, the time development operator and the Green's operators can depend only upon the relative time,

$$U(t_1\,,\,t_2) = U(t) \;, \quad G(t_1\,,\,t_2) = G(t), \quad t = t_1 - t_2 \quad , \tag{8.15}$$

and the defining properties of the Green's operators appear as

$$\left(i \frac{\partial}{\partial t} - H\right) G(t) = G(t) \left(i \frac{\partial^T}{\partial t} - H\right) = \delta(t) \tag{8.16}$$

$$G_r(t) = 0 \ , \quad t < 0 \quad ; \qquad G_a(t) = G_r(-t)^\dagger = 0 \ , \quad t > 0 \ .$$

It is now possible to eliminate the time dependence in the Green's operators by defining the transform operators

$$G_r(E)$$

$$= \int_{-\infty}^{\infty} dt \; e^{iEt} \, G_r(t) = \int_0^{\infty} dt \; e^{iEt} \, G_r(t) \ , \quad \text{Im } E > 0$$

$$G_a(E) \tag{8.17}$$

$$= \int_{-\infty}^{\infty} dt \; e^{iEt} \, G_a(t) = \int_{-\infty}^{0} dt \; e^{iEt} \, G_a(t) \ , \quad \text{Im } E < 0 \ .$$

As we have already indicated, since the time integrations are extended only over semi-infinite intervals these operators exist for complex values of the energy parameter E , when restricted to the appropriate half-plane. The application of the transformation to the differential equations (8.16) yields

$$(E-H) \; G(E) = G(E) \; (E-H) = 1 \tag{8.18}$$

for both Green's operators which, as functions of the complex variable E, are now defined by the respective domains of regularity indicated in (8.17). Correspondingly the adjoint connection now appears as

$$G_r(E)^\dagger = G_a(E^*) \quad . \tag{8.19}$$

The inversions of (8.17) are comprised in

$$G(t) = \frac{1}{2\pi} \int_{-\infty}^{\infty} dE \; e^{-iEt} \; G(E) \tag{8.20}$$

where the integration path, extended parallel to the real axis, is drawn in the domain of regularity appropriate to the Green's operator under consideration.

8.3 OPERATOR FUNCTION OF A COMPLEX VARIABLE

Both Green's operators are given formally by

$$G(E) = \frac{1}{E-H} = \sum_{E'\gamma'} \frac{|E'\gamma'\rangle\langle E'\gamma'|}{E - E'} \tag{8.21}$$

and therefore form together a single operator function of the complex variable E, defined everywhere except perhaps on the common boundary of the two half-planes, the real axis. Indeed, the form (8.21), expressed in terms of the eigenvectors of the operator H and a supplementary set of constants of the motion γ, shows that the singularities of $G(E)$ are simple poles on the real E axis, coinciding with the spectrum of energy values for the system. The construction of the Green's function in some convenient representation, and an investigation of its singularities will thus supply the entire energy spectrum of the system together with automatically normalized and complete sets of wave functions for the energy states. For a system with the Hamiltonian $H(q^{\dagger}, q)$, for example, the Green's function $G(q^{\dagger\prime}, q', E)$ could be obtained by solving the inhomogeneous differential equation

$$\left[E - H\left(q^{\dagger\prime}, \frac{\partial_{\ell}}{\partial q^{\dagger\prime}}\right)\right] G(q^{\dagger\prime}, q', E) \tag{8.22}$$
$$= \langle q^{\dagger\prime} | q' \rangle = \exp\left(\sum_a q^{\dagger\prime}_a q'_a\right) .$$

On exhibiting the solution as

$$G(q^{\dagger\prime}, q', E) = \sum_{E'\gamma'} \frac{\langle q^{\dagger\prime} | E'\gamma' \rangle \langle E'\gamma' | q' \rangle}{E - E'} \tag{8.23}$$

the desired information concerning energy values and wavefunctions is disclosed. We should also note the possibility of a partial Green's function construction, which supplies information about a selected group of states. Thus if we place the eigenvalues q' equal to zero in (8.22), the differential equation reads

$$\left[E - H\left(q^{\dagger\prime}, \frac{\partial_\ell}{\partial q^{\dagger\prime}}\right)\right] G\left(q^{\dagger\prime}, 0, E\right) = 1 \qquad (8.24)$$

and the solution of this equation will yield the energy values only for those states with $\langle E'\gamma'|0\rangle \neq 0$. All these states are still represented in the further specialized Green's function

$$G(0, 0, E) = \sum_{E'\gamma'} \frac{|\langle E'\gamma'|0\rangle|^2}{E - E'} \qquad (8.25)$$

where the coefficients $|\langle E'\gamma'|0\rangle|^2$ obey

$$\sum_{E'\gamma'} |\langle E'\gamma'|0\rangle|^2 = 1 , \qquad (8.26)$$

and give the probabilities for realizing the various energy states in a measurement on the zero eigenvalue state $|0\rangle$.

8.4 SINGULARITIES

The singularities of a Green's function can be determined from the discontinuities encountered on crossing the real E axis. Thus, for real E,

$$\lim_{\varepsilon\to+0} [G(E+i\varepsilon) - G(E-i\varepsilon)]$$

$$= \lim_{\varepsilon\to+0} \left[\frac{1}{E-H+i\varepsilon} - \frac{1}{E-H-i\varepsilon} \right] = -2\pi i\ \delta(E-H) \qquad (8.27)$$

$$= -2\pi i \sum_{E'\gamma'} \delta(E-E')\ |E'\gamma'\rangle\langle E'\gamma'| \quad ,$$

according to the delta function construction

$$\delta(z) = \lim_{\varepsilon\to+0} \frac{1}{\pi} \frac{\varepsilon}{z^2+\varepsilon^2} \quad . \qquad (8.28)$$

Note that this is also a measure of the non-Hermitian part of $G_r(E)$ for real E,

$$\lim_{\varepsilon\to+0} [G(E+i\varepsilon) - G(E-i\varepsilon)] = G_r(E) - G_r(E)^{\dagger} \quad . \qquad (8.29)$$

There is no discontinuity and G_r is Hermitian at any point on the real axis that does not belong to the energy spectrum of the system, whereas a discrete energy value is recognized by the corresponding localized discontinuity. If the energy spectrum forms a continuum beginning at E_0, the discontinuity for $E > E_0$ is

$$-2\pi i \sum_{\gamma'} \int_{E_0}^{\infty} dE' \, \delta(E-E') \, |E'\gamma'><E'\gamma'|$$
$$= -2\pi i \sum_{\gamma'} |E\gamma'><E\gamma'| \quad . \tag{8.30}$$

The existence of such a finite discontinuity for every $E > E_0$ implies that $G(E)$ possesses a branch point singularity at $E = E_0$, the continuous line of poles extending from E_0 to infinity supplying the cut in the E plane. Thus the precise nature of the energy spectrum for a system is implied by the character of the singularities exhibited by $G(E)$ as a function of a complex variable.

8.5 AN EXAMPLE

The elementary example of a single free particle in space is described non-relativistically in the $\mathbf{r}$ representation by the Green's function equation

$$\left(E + \frac{1}{2m}\nabla^2\right) G(\mathbf{r}, \mathbf{r}', E) = \delta(\mathbf{r}-\mathbf{r}') \quad . \tag{8.31}$$

The solution

$$G(\mathbf{r}, \mathbf{r}', E) = -\frac{m}{2\pi} \frac{e^{ip|\mathbf{r}-\mathbf{r}'|}}{|\mathbf{r}-\mathbf{r}'|} \quad , \quad p = (2mE)^{\frac{1}{2}} \tag{8.32}$$

involves the double-valued function $E^{\frac{1}{2}}$ of the complex energy parameter, which must be interpreted as $+i|E|^{\frac{1}{2}}$ for $E < 0$ in order that the Green's function remain bounded as $|\mathbf{r}-\mathbf{r}'| \to \infty$. Accordingly, for $E > 0$ we must have $E^{\frac{1}{2}} = +|E|^{\frac{1}{2}}$ on the upper half of the real axis and $E^{\frac{1}{2}} = -|E|^{\frac{1}{2}}$ on the lower half. Hence there is a discontinuity across the real axis for $E > 0$ which constitutes the entire energy spectrum, since the Green's function is always bounded as a function of E . The discontinuity is given by

$$\begin{aligned} & G_r(\mathbf{r}, \mathbf{r}', E) - G_a(\mathbf{r}, \mathbf{r}', E) \\ &= 2i \operatorname{Im} G_r(\mathbf{r}, \mathbf{r}', E) \\ &= -i \frac{m}{\pi} \frac{\sin p|\mathbf{r}-\mathbf{r}'|}{|\mathbf{r}-\mathbf{r}'|} \quad , \quad p = +|p| \qquad (8.33) \\ &= -2\pi i \int d\omega \, \frac{mp}{(2\pi)^3} e^{i\mathbf{p}\cdot(\mathbf{r}-\mathbf{r}')} \end{aligned}$$

where the integral in the latter form is extended over all directions of the vector $\mathbf{p} = p\,\mathbf{n}$. On comparison with (8.30) we see that the various states of a given energy can be labelled by the unit vector $\mathbf{n}$, specified within an infinitesimal solid angle $d\omega$. The corresponding wave functions of the $\mathbf{r}$ representation are

$$\langle \mathbf{r}|E\mathbf{n}\rangle = \left[d\omega \, \frac{mp}{(2\pi)^3} \right]^{\frac{1}{2}} e^{i\mathbf{p}\cdot\mathbf{r}} = \left[\frac{(d\mathbf{p})}{(2\pi)^3} \, \frac{1}{dE} \right]^{\frac{1}{2}} e^{i\mathbf{p}\cdot\mathbf{r}} \tag{8.34}$$

where $(d\mathbf{p})$ is the element of volume in the p space.

8.6 PARTIAL GREEN'S FUNCTION

The utility of a partial Green's function construction appears in two general situations, which can overlap. One or more compatible constants of the motion may be apparent from symmetry considerations and it is desired to investigate only states with specific values of these quantities, or, one may be interested for classification purposes in constructing the states of a perturbed system which correspond most closely to certain states of a related unperturbed system. Both situations can be characterized as a decomposition of the complete set of states into two parts, or subspaces, as symbolized by

$$1 = M_1 + M_2 \quad , \tag{8.35}$$

in which the measurement symbols, or projection operators, obey

$$M_1{}^2 = M_1 \quad , \quad M_2{}^2 = M_2 \quad , \quad M_1 M_2 = M_2 M_1 = 0 \quad , \tag{8.36}$$

and where one seeks to construct the projected Green's operator referring only to the subspace M_1 ,

$$M_1\, G(E) M_1 = {}_1G_1(E) \quad . \tag{8.37}$$

From the equation

$$(E-H)G(E) = 1 \tag{8.18}$$

one obtains

$$M_1(E-H)\ \left(M_1+M_2\right)G(E)\ M_1 = M_1 \tag{8.38}$$

and

$$M_2(E-H)\left(M_1+M_2\right)G(E)\,M_1 = 0 \quad , \tag{8.39}$$

which we write as

$$\begin{aligned} \left(E - {}_1H_1\right)\, {}_1G_1(E) - {}_1H_2\, {}_2G_1(E) &= M_1 \\ \left(E - {}_2H_2\right)\, {}_2G_1(E) - {}_2H_1\, {}_1G_1(E) &= 0 \quad . \end{aligned} \tag{8.40}$$

The second equation is formally solved, within the subspace M_2 , by

$${}_2G_1(E) = \frac{1}{E - {}_2H_2}\, {}_2H_1\, {}_1G_1(E) \quad , \tag{8.41}$$

and we obtain as the determining equation for ${}_1G_1(E)$,

$$[E - {}_1H_1 - \Delta_1H_1(E)]\, {}_1G_1(E) = M_1 \tag{8.42}$$

where

$$\Delta_1H_1 = {}_1H_2\, \frac{1}{E - {}_2H_2}\, {}_2H_1 \quad . \tag{8.43}$$

If the two subspaces refer to distinct values of constants of the motion for the complete Hamiltonian, there will be no matrix elements connecting the subspaces, ${}_1H_2 = {}_2H_1 = 0$, and (8.42) reduces to the fundamental form of the Green's operator equation,

(8.18), now defined entirely within the space M_1 .

CHAPTER NINE

SOME APPLICATIONS AND FURTHER DEVELOPMENTS

9.1 BROWNIAN MOTION OF A QUANTUM OSCILLATOR†

---

† Reproduced from the Journal of Mathematical Physics, Vol. 2, pp. 407 -432 (1961).

Reprinted from the JOURNAL OF MATHEMATICAL PHYSICS, Vol. 2, No. 3, 407–432, May–June, 1961
Printed in U. S. A.

Brownian Motion of a Quantum Oscillator

JULIAN SCHWINGER*
Harvard University, Cambridge, Massachusetts
(Received November 28, 1960)

An action principle technique for the direct computation of expectation values is described and illustrated in detail by a special physical example, the effect on an oscillator of another physical system. This simple problem has the advantage of combining immediate physical applicability (e.g., resistive damping or maser amplification of a single electromagnetic cavity mode) with a significant idealization of the complex problems encountered in many-particle and relativistic field theory. Successive sections contain discussions of the oscillator subjected to external forces, the oscillator loosely coupled to the external system, an improved treatment of this problem and, finally, there is a brief account of a general formulation.

INTRODUCTION

THE title of this paper refers to an elementary physical example that we shall use to illustrate, at some length, a solution of the following methodological problem. The quantum action principle[1] is a differential characterization of transformation functions, $\langle a't_1|b't_2\rangle$, and thus is ideally suited to the practical computation of transition probabilities (which includes the determination of stationary states). Many physical questions do not pertain to individual transition probabilities, however, but rather to expectation values of a physical property for a specified initial state,

$$\langle X(t_1)\rangle_{b't_2}=\sum_{a'a''}\langle b't_2|a't_1\rangle\langle a't_1|X(t_1)|a''t_1\rangle\langle a''t_1|b't_2\rangle,$$

or, more generally, a mixture of states. Can one devise an action principle technique that is adapted to the direct computation of such expectation values, without requiring knowledge of the individual transformation functions?

The action principle asserts that ($\hbar=1$),

$$\delta\langle a't_1|b't_2\rangle=i\left\langle a't_1\left|\delta\left[\int_{t_2}^{t_1}dtL\right]\right|b't_2\right\rangle,$$

and

$$\delta\langle b't_2|a't_1\rangle=-i\left\langle b't_2\left|\delta\left[\int_{t_2}^{t_1}dtL\right]\right|a't_1\right\rangle,$$

in which we shall take $t_1>t_2$. These mutually complex-conjugate forms correspond to the two viewpoints whereby states at different times can be compared, either by progressing forward from the earlier time, or backward from the later time. The relation between the pair of transformation functions is such that

$$\delta\Big[\sum_{a'}\langle b't_2|a't_1\rangle\langle a't_1|b''t_2\rangle\Big]=0,$$

which expresses the fixed numerical value of

$$\langle b't_2|b''t_2\rangle=\delta(b',b'').$$

But now, imagine that the positive and negative senses of time development are governed by different dynamics. Then the transformation function for the closed circuit will be described by the action principle

$$\delta\langle t_2|t_2\rangle=\delta[\langle t_2|t_1\rangle\times\langle t_1|t_2\rangle]$$
$$=i\left\langle t_2\left|\delta\left[\int_{t_2}^{t_1}dtL_+-\int_{t_2}^{t_1}dtL_-\right]\right|t_2\right\rangle,$$

in which abbreviated notation the multiplication sign symbolizes the composition of transformation functions by summation over a complete set of states. If, in particular, the Lagrangian operators $L_\pm$ contain the dynamical term $\lambda_\pm(t)X(t)$, we have

$$\delta_\lambda\langle t_2|t_2\rangle=i\left\langle t_2\left|\int_{t_2}^{t_1}dt(\delta\lambda_+-\delta\lambda_-)X(t)\right|t_2\right\rangle,$$

and, therefore,

$$-i\frac{\delta}{\delta\lambda_+(t_1)}\langle t_2|t_2\rangle=i\frac{\delta}{\delta\lambda_-(t_1)}\langle t_2|t_2\rangle$$
$$=\langle t_2|X(t_1)|t_2\rangle,$$

where $\lambda_\pm$ can now be identified. Accordingly, if a system is suitably perturbed[2] in a manner that depends upon the time sense, a knowledge of the transformation function referring to a closed time path determines the expectation value of any desired physical quantity for a specified initial state or state mixture.

OSCILLATOR

To illustrate this remark we first consider an oscillator subjected to an arbitrary external force, as described by the Lagrangian operator

$$L=iy^\dagger(dy/dt)-\omega y^\dagger y-y^\dagger K(t)-yK^*(t),$$

* Supported by the Air Force Office of Scientific Research (ARDC).

[1] Some references are: Julian Schwinger, Phys. Rev. **82**, 914 (1951); **91**, 713 (1953); Phil. Mag. **44**, 1171 (1953). The first two papers also appear in *Selected Papers on Quantum Electrodynamics* (Dover Publications, New York, 1958). A recent discussion is contained in Julian Schwinger, Proc. Natl. Acad. Sci. U. S. **46**, 883 (1960).

[2] Despite this dynamical language, a change in the Hamiltonian operator of a system can be kinematical in character, arising from the consideration of another transformation along with the dynamical one generated by the Hamiltonian. See the last paper quoted in footnote 1, and Julian Schwinger, Proc. Natl. Acad. Sci. U. S. **46**, 1401 (1960).

in which the complementary pair of non-Hermitian operators y, $iy^\dagger$, are constructed from Hermitian operators q, p by

$$y=2^{-\frac{1}{2}}(q+ip)$$
$$iy^\dagger=2^{-\frac{1}{2}}(p+iq).$$

The equations of motion implied by the action principle are

$$i(dy/dt)-\omega y=K$$
$$-i(dy^\dagger/dt)-\omega y^\dagger=K^*,$$

and solutions are given by

$$y(t)=e^{-i\omega(t-t_2)}y(t_2)-i\int_{t_2}^{t}dt'e^{-i\omega(t-t')}K(t'),$$

together with the adjoint equation. Since we now distinguish between the forces encountered in the positive time sense, $K_+(t)$, $K_+^*(t)$, and in the reverse time direction, $K_-(t)$, $K_-^*(t)$, the integral must be taken along the appropriate path. Thus, when t is reached first in the time evolution from t_2, we have

$$y_+(t)=e^{-i\omega(t-t_2)}y_+(t_2)-i\int_{t_2}^{t}dt'e^{-i\omega(t-t')}K_+(t'),$$

while on the subsequent return to time t,

$$y_-(t)=e^{-i\omega(t-t_2)}y_+(t_2)-i\int_{t_2}^{t_1}dt'e^{-i\omega(t-t')}K_+(t')$$
$$+i\int_{t}^{t_1}dt'e^{-i\omega(t-t')}K_-(t').$$

Note that

$$y_-(t_1)-y_+(t_1)=0,$$
$$y_-(t_2)-y_+(t_2)=i\int_{t_2}^{t_1}dte^{i\omega(t-t_2)}(K_--K_+)(t).$$

We shall begin by constructing the transformation function referring to the lowest energy state of the unperturbed oscillator, $\langle 0t_2|0t_2\rangle^{K\pm}$. This state can be characterized by

$$\langle 0t_2|y^\dagger y(t_2)|0t_2\rangle=0$$

or, equivalently, by the eigenvector equations

$$y(t_2)|0t_2\rangle=0,\quad \langle 0t_2|y^\dagger(t_2)=0.$$

Since the transformation function simply equals unity if $K_+=K_-$ and $K_+^*=K_-^*$, we must examine the effect of independent changes in K_+ and K_-, and of K_+^* and K_-^*, as described by the action principle

$$\delta_K\langle 0t_2|0t_2\rangle^{K\pm}=-i\left\langle 0t_2\left|\left[\int_{t_2}^{t_1}dt(\delta K_+^*y_+-\delta K_-^*y_-)+\int_{t_2}^{t_1}dt(y_+^\dagger\delta K_+-y_-^\dagger\delta K_-)\right]\right|0t_2\right\rangle^{K\pm}.$$

The choice of initial state implies effective boundary conditions that supplement the equations of motion,

$$y_+(t_2)\to 0,\quad y_-^\dagger(t_2)\to 0.$$

Hence, in effect we have

$$y_+(t)=-i\int_{t_2}^{t_1}dt'e^{-i\omega(t-t')}\eta_+(t-t')K_+(t')$$

and

$$y_-(t)=-i\int_{t_2}^{t_1}dt'e^{-i\omega(t-t')}K_+(t')$$
$$+i\int_{t_2}^{t_1}dt'e^{-i\omega(t-t')}\eta_-(t-t')K_-(t'),$$

together with the similar adjoint equations obtained by interchanging the $\pm$ labels. For convenience, step functions have been introduced:

$$\eta_+(t-t')=\begin{cases}1, & t-t'>0\\ 0, & t-t'<0\end{cases},$$

$$\eta_-(t-t')=\begin{cases}1, & t-t'<0\\ 0, & t-t'>0\end{cases},$$

$$\eta_+(t-t')+\eta_-(t-t')=1,\quad \eta_+(0)=\eta_-(0)=\tfrac{1}{2}.$$

We shall also have occasion to use the odd function

$$\epsilon(t-t')=\eta_+(t-t')-\eta_-(t-t').$$

The solution of the resulting integrable differential expression for $\log\langle 0t_2|0t_2\rangle^{K\pm}$ is given by

$$\langle 0t_2|0t_2\rangle^{K\pm}=\exp\left[-i\int_{t_2}^{t_1}dtdt'K^*(t)G_0(t-t')K(t')\right],$$

in a matrix notation, with

$$K(t)=\begin{pmatrix}K_+(t)\\ K_-(t)\end{pmatrix}$$

and

$$iG_0(t-t')=e^{-i\omega(t-t')}\begin{pmatrix}\eta_+(t-t') & 0\\ -1 & \eta_-(t-t')\end{pmatrix}.$$

The requirement that the transformation function reduce to unity on identifying K_+ with K_-, K_+^* with K_-^*, is satisfied by the null sum of all elements of G_0, as assured by the property $\eta_++\eta_-=1$.

An operator interpretation of G_0 is given by the second variation

$$-\delta_{K^*}\delta_K\langle 0t_2|0t_2\rangle^{K\pm}|_{K=K^*=0}$$
$$=i\int dtdt'\delta K^*(t)G_0(t-t')\delta K(t').$$

Generally, on performing two distinct variations in the structure of L that refer to parameters upon which the dynamical variables at a given time are not explicitly dependent, we have

$$-\delta_1\delta_2\langle t_2|t_2\rangle = \left\langle t_2 \left| \int_{t_2}^{t_1} dt dt' \{ (\delta_1 L_+(t)\delta_2 L_+(t'))_+ - \delta_1 L_-(t)\delta_2 L_+(t') - \delta_2 L_-(t)\delta_1 L_+(t') + (\delta_1 L_-(t)\delta_2 L_-(t'))_- \} \right| t_2 \right\rangle,$$

in which the multiplication order follows the sense of time development. Accordingly,

$$iG_0(t-t') = \begin{pmatrix} \langle (y(t)y^\dagger(t'))_+\rangle_0 & -\langle y^\dagger(t')y(t)\rangle_0 \\ -\langle y(t)y^\dagger(t')\rangle_0 & \langle (y(t)y^\dagger(t'))_-\rangle_0 \end{pmatrix},$$

where the expectation values and operators refer to the lowest state and the dynamical variables of the unperturbed oscillator. The property of G_0 that the sum of rows and columns vanishes is here a consequence of the algebraic property

$$(y(t)y^\dagger(t'))_+ + (y(t)y^\dagger(t'))_- = \{y(t), y^\dagger(t')\}.$$

The choice of oscillator ground state is no essential restriction since we can now derive the analogous results for any initial oscillator state. Consider, for this purpose, the impulse forces

$$K_+(t) = iy''\delta(t-t_2),$$
$$K_-^*(t) = -iy^{\dagger\prime}\delta(t-t_2),$$

the effects of which are described by

$$y_+(t_2+0) - y_+(t_2) = y'',$$
$$y_-^\dagger(t_2+0) - y_-^\dagger(t_2) = y^{\dagger\prime}.$$

Thus, under the influence of these forces, the states $|0t_2\rangle$ and $\langle 0t_2|$ become, at the time t_2+0, the states $|y''t_2\rangle$ and $\langle y^{\dagger\prime}t_2|$, which are right and left eigenvectors, respectively, of the operators $y(t_2)$ and $y^\dagger(t_2)$. On taking into account arbitrary additional forces, the transformation function for the closed time path can be expressed as

$$\langle y^{\dagger\prime}t_1|y''t_2\rangle^{K\pm} = \exp\left[y^{\dagger\prime}y'' - y^{\dagger\prime}\left(\int_{t_2}^{t_1} dt G_0(t_2-t)K(t)\right)_- + \left(\int_{t_2}^{t_1} dt K^*(t)G_0(t-t_2)\right)_+ y'' - i\int_{t_2}^{t_1} dt dt' K^*(t)G_0(t-t')K(t') \right],$$

in which

$$\left(\int_{t_2}^{t_1} dt G_0(t_2-t)K(t)\right)_- = -i\int_{t_2}^{t_1} dt e^{i\omega(t-t_2)}(K_- - K_+)(t)$$

and

$$\left(\int dt K^*(t)G_0(t-t_2)\right)_+ = -i\int_{t_2}^{t_1} dt e^{-i\omega(t-t_2)}(K_+^* - K_-^*)(t).$$

The eigenvectors of the non-Hermitian canonical variables are complete and have an intrinsic physical interpretation in terms of q and p measurements of optimum compatibility.[3] For our immediate purposes, however, we are more interested in the unperturbed oscillator energy states. The connection between the two descriptions can be obtained by considering the unperturbed oscillator transformation function

$$\langle y^{\dagger\prime}t_1|y''t_2\rangle = \langle y^{\dagger\prime}|\exp[-i(t_1-t_2)\omega y^\dagger y]|y''\rangle.$$

Now

$$i(\partial/\partial t_1)\langle y^{\dagger\prime}t_1|y''t_2\rangle = \langle y^{\dagger\prime}t_1|\omega y^\dagger(t_1)y(t_1)|y''t_2\rangle = \omega y^{\dagger\prime}e^{-i\omega(t_1-t_2)}y''\langle y^{\dagger\prime}t_1|y''t_2\rangle,$$

since

$$y(t_1) = e^{-i\omega(t_1-t_2)}y(t_2),$$

which yields

$$\langle y^{\dagger\prime}t_1|y''t_2\rangle = \exp[y^{\dagger\prime}e^{-i\omega(t_1-t_2)}y''] = \sum_{n=0}^{\infty} \frac{(y^{\dagger\prime})^n}{(n!)^{\frac{1}{2}}} e^{-in\omega(t_1-t_2)} \frac{(y'')^n}{(n!)^{\frac{1}{2}}}.$$

We infer the nonnegative integer spectrum of $y^\dagger y$, and the corresponding wave functions

$$\langle y^{\dagger\prime}|n\rangle = (y^{\dagger\prime})^n/(n!)^{\frac{1}{2}}, \quad \langle n|y'\rangle = (y')^n/(n!)^{\frac{1}{2}}.$$

Accordingly, a non-Hermitian canonical variable transformation function can serve as a generator for the transformation function referring to unperturbed oscillator energy states,

$$\langle y^{\dagger\prime}t_2|y''t_2\rangle^{K\pm} = \sum_{n,n'=0}^{\infty} \frac{(y^{\dagger\prime})^n}{(n!)^{\frac{1}{2}}} \langle nt_2|n't_2\rangle^{K\pm} \frac{(y'')^{n'}}{(n'!)^{\frac{1}{2}}}.$$

If we are specifically interested in $\langle nt_2|nt_2\rangle^{K\pm}$, which supplies all expectation values referring to the initial state n, we must extract the coefficient of $(y^{\dagger\prime}y'')^n/n!$ from an exponential of the form

$$\exp[y^{\dagger\prime}y'' + y^{\dagger\prime}\alpha + \beta y'' + \gamma] = \sum_{kl} \frac{(y^{\dagger\prime})^k}{k!} \frac{(y'')^l}{l!} \alpha^k\beta^l \exp[y^{\dagger\prime}y'' + \gamma].$$

All the terms that contribute to the required coefficient

[3] A discussion of non-Hermitian representations is given in *Lectures on Quantum Mechanics* (Les Houches, 1955), unpublished.

are contained in

$$\sum_{k=0}^{\infty} \frac{(y^{\dagger\prime}y^{\prime\prime})^k}{(k!)^2}(\alpha\beta)^k \exp[y^{\dagger\prime}y^{\prime\prime}+\gamma]$$
$$=\frac{1}{2\pi i}\oint \frac{d\lambda}{\lambda}e^{\lambda}\exp[y^{\dagger\prime}y^{\prime\prime}(1+\lambda^{-1}\alpha\beta)+\gamma],$$

where the latter version is obtained from

$$\frac{1}{2\pi i}\oint \frac{d\lambda}{\lambda^{k+1}}e^{\lambda}=\frac{1}{k!},$$

and

$$\langle nt_2|nt_2\rangle^{K\pm}=\exp\left[-i\int dtdt' K^*(t)G_0(t-t')K(t')\right]$$
$$\times L_n\left[\left(\int dtK^*(t)G_0(t-t_2)\right)_+ \times\left(\int dtG_0(t_2-t)K(t)\right)_-\right],$$

in which the nth Laguerre polynomial has been introduced on observing that

$$\frac{1}{2\pi i}\oint \frac{d\lambda}{\lambda}e^{\lambda}(1-\lambda^{-1}x)^n=\frac{1}{n!}e^x\left(\frac{d}{dx}\right)^n x^ne^{-x}=L_n(x).$$

One obtains a much neater form, however, from which these results can be recovered, on considering an initial mixture of oscillator energy states for which the nth state is assigned the probability

$$(1-e^{-\beta\omega})e^{-n\beta\omega},$$

and

$$\beta^{-1}=\vartheta$$

can be interpreted as a temperature. Then, since

$$(1-e^{-\beta\omega})\sum_{n=0}^{\infty} e^{-n\beta\omega}L_n(x)=(1-e^{-\beta\omega})\frac{1}{2\pi i}\oint\frac{d\lambda}{\lambda}$$
$$\times e^{\lambda}[1-e^{-\beta\omega}+\lambda^{-1}e^{-\beta\omega}x]^{-1}=\exp\left[-\frac{x}{e^{\beta\omega}-1}\right],$$

we obtain

$$\langle t_2|t_2\rangle_\vartheta{}^{K\pm}=\exp\left[-i\int_{t_2}^{t_1} dtdt' K^*(t)G_\vartheta(t-t')K(t')\right],$$

with

$$iG_\vartheta(t-t')=iG_0(t-t')+(e^{\beta\omega}-1)^{-1}G_0(t-t_2)_+\,{}_-G_0(t_2-t'),$$

and in which

$$iG_0(t-t_2)_+=e^{-i\omega(t-t_2)}\begin{pmatrix}1\\-1\end{pmatrix},$$

$$i_-G_0(t_2-t)=e^{i\omega(t-t_2)}\begin{pmatrix}-1 & 1\end{pmatrix}.$$

Thus,

$$iG_\vartheta(t-t')=e^{-i\omega(t-t')}\begin{pmatrix}\eta_+(t-t')+\langle n\rangle_\vartheta, & -\langle n\rangle_\vartheta\\ -1-\langle n\rangle_\vartheta, & \eta_-(t-t')+\langle n\rangle_\vartheta\end{pmatrix},$$

where we have written

$$\langle n\rangle_\vartheta=(e^{\beta\omega}-1)^{-1},$$

and since the elements of G_ϑ are also given by unperturbed oscillator thermal expectation values

$$iG_\vartheta(t-t')=\begin{pmatrix}\langle(y(t)y^\dagger(t'))_+\rangle_\vartheta & -\langle y^\dagger(t')y(t)\rangle_\vartheta\\ -\langle y(t)y^\dagger(t')\rangle_\vartheta & \langle(y(t)y^\dagger(t'))_-\rangle_\vartheta\end{pmatrix},$$

the designation $\langle n\rangle_\vartheta$ is consistent with its identification as $\langle y^\dagger y\rangle_\vartheta$.

The thermal forms can also be derived directly by solving the equations of motion, in the manner used to find $\langle 0t_2|0t_2\rangle^{K\pm}$. On replacing the single diagonal element

$$\langle 0t_2|0t_2\rangle^{K\pm}=\langle 0t_2|U|0t_2\rangle$$

by the statistical average

$$(1-e^{-\beta\omega})\sum_0^\infty e^{-n\beta\omega}\langle nt_2|nt_2\rangle^{K\pm}$$
$$=(1-e^{-\beta\omega})\,\mathrm{tr}[\exp(-\beta\omega y^\dagger y)U],$$

we find the following relation,

$$y_-(t_2)=e^{\beta\omega}y_+(t_2),$$

instead of the effective initial condition $y_+(t_2)=0$. This is obtained by combining

$$\exp(-\beta\omega y^\dagger y)y\exp(\beta\omega y^\dagger y)=\exp(\beta\omega)y$$

with the property of the trace

$$\mathrm{tr}[\exp(-\beta\omega y^\dagger y)yU]=\mathrm{tr}[\exp(\beta\omega)y\exp(-\beta\omega y^\dagger y)U]$$
$$=\mathrm{tr}[\exp(-\beta\omega y^\dagger y)U\exp(\beta\omega)y].$$

We also have

$$y_-(t_2)-y_+(t_2)=-i\int_{t_2}^{t_1} dte^{i\omega(t-t_2)}(K_+-K_-)(t),$$

and therefore, effectively,

$$y_+(t_2)=-i\frac{1}{e^{\beta\omega}-1}\int_{t_2}^{t_1} dte^{i\omega(t-t_2)}(K_+-K_-)(t).$$

Hence, to the previously determined $y_\pm(t)$ is to be added the term

$$-i\langle n\rangle_\vartheta\int_{t_2}^{t_1} dt'e^{-i\omega(t-t')}(K_+-K_-)(t'),$$

and correspondingly

$$\langle t_2|t_2\rangle_\vartheta{}^{K\pm}=\langle t_2|t_2\rangle_0{}^{K\pm}\exp\left[-\langle n\rangle_\vartheta\int_{t_2}^{t_1} dtdt'\right.$$
$$\left.\times(K_+{}^*-K_-{}^*)(t)e^{-i\omega(t-t')}(K_+-K_-)(t')\right],$$

which reproduces the earlier result.

As an elementary application let us evaluate the expectation value of the oscillator energy at time t_1 for a system that was in thermal equilibrium at time t_2 and is subsequently disturbed by an arbitrarily time-varying force. This can be computed as

$$\langle t_2|\omega y^\dagger y(t_1)|t_2\rangle_\vartheta{}^K = \omega\frac{\delta}{\delta K_-(t_1)}\frac{\delta}{\delta K_+{}^*(t_1)}\langle t_2|t_2\rangle_\vartheta{}^{K\pm}|_{K_+=K_-,\,K_+{}^*=K_-{}^*}.$$

The derivative $\delta/\delta K_+{}^*(t_1)$ supplies the factor

$$-i\left(\int_{t_2}^{t_1} dt G_\vartheta(t_1-t')K(t')\right)_+,$$

the subsequent variation with respect to $K_-(t_1)$ gives

$$-iG_\vartheta(0)_{+-}+\left(\int dt K^*(t)G_\vartheta(t-t_1)\right)_- \times\left(\int dt' G_\vartheta(t_1-t')K(t')\right)_+,$$

and the required energy expectation value equals

$$\omega\langle n\rangle_\vartheta+\omega\left|\int_{t_2}^{t_1} dt e^{i\omega t}K(t)\right|^2.$$

More generally, the expectation values of all functions of $y(t_1)$ and $y^\dagger(t_1)$ are known from that of

$$\exp\{-i[\lambda y^\dagger(t_1)+\mu y(t_1)]\},$$

and this quantity is obtained on supplementing K_+ and $K_+{}^*$ by the impulsive forces (note that in this use of the formalism a literal complex-conjugate relationship is not required)

$$K_+(t)=\lambda\delta(t-t_1),$$
$$K_+{}^*(t)=\mu\delta(t-t_1).$$

Then

$$\langle t_2|\exp\{-i[\lambda y^\dagger(t_1)+\mu y(t_1)]\}|t_2\rangle_\vartheta{}^K = \exp\left[-\lambda\mu(\langle n\rangle_\vartheta+\tfrac{1}{2})+\lambda\int_{t_2}^{t_1} dt e^{i\omega(t_1-t)}K^*(t) -\mu\int_{t_2}^{t_1} dt e^{-i\omega(t_1-t)}K(t)\right],$$

which involves the special step-function value

$$\eta_+(0)=\tfrac{1}{2}.$$

Alternatively, if we choose

$$K_+(t)=\lambda\delta(t-t_1),$$
$$K_+{}^*(t)=\mu\delta(t-t_1+0),$$

there appears

$$\langle t_2|\exp[-i\lambda y^\dagger(t_1)]\exp[-i\mu y(t_1)]|t_2\rangle_\vartheta{}^K = \exp\left[-\lambda\mu\langle n\rangle_\vartheta+\lambda\int_{t_2}^{t_1} dt e^{i\omega(t_1-t)}K^*(t) -\mu\int_{t_2}^{t_1} dt e^{-i\omega(t_1-t)}K(t)\right].$$

It may be worth remarking, in connection with these results, that the attention to expectation values does not deprive us of the ability to compute individual probabilities. Indeed, if probabilities for specific oscillator energy states are of interest, we have only to exhibit, as functions of y and $y^\dagger$, the projection operators for these states, the expectation values of which are the required probabilities. Now

$$P_n=|n\rangle\langle n|$$

is represented by the matrix

$$\langle y^{\dagger\prime}|P_n|y''\rangle=(y^{\dagger\prime}y'')^n/n! = [(y^{\dagger\prime}y'')^n/n!]\exp(-y^{\dagger\prime}y'')\langle y^{\dagger\prime}|y''\rangle,$$

and, therefore,

$$P_n=\frac{1}{n!}(y^\dagger)^n\left[\sum_{k=0}^{\infty}\frac{(-1)^k}{k!}(y^\dagger)^k y^k\right]y^n = \frac{1}{n!}(y^\dagger)^n\exp(-y^\dagger;y)y^n,$$

in which we have introduced a notation to indicate this ordered multiplication of operators. A convenient generating function for these projection operators is

$$\sum_{n=0}^{\infty}\alpha^n P_n=\exp[-(1-\alpha)y^\dagger;y],$$

and we observe that

$$\sum_0^\infty \alpha^n P_n=\exp\left[(1-\alpha)\frac{\partial}{\partial\lambda}\frac{\partial}{\partial\mu}\right]\times\exp(-i\lambda y^\dagger)\exp(-i\mu y)|_{\lambda=\mu=0}.$$

Accordingly,

$$\sum_0^\infty \alpha^n p(n,\vartheta,K)=\exp\left[(1-\alpha)\frac{\partial}{\partial\lambda}\frac{\partial}{\partial\mu}\right]\exp\left[-\lambda\mu\langle n\rangle_\vartheta +\lambda e^{i\omega t_1}\int dt e^{-i\omega t}K^*(t) -\mu e^{-i\omega t_1}\int dt e^{i\omega t}K(t)\right]\Bigg|_{\lambda=\mu=0}$$

gives the probability of finding the oscillator in the nth energy state after an arbitrary time-varying force has acted, if it was initially in a thermal mixture of states.

To evaluate

$$X=\exp\left[(1-\alpha)\frac{\partial}{\partial\lambda}\frac{\partial}{\partial\mu}\right]\exp[-\lambda\mu\langle n\rangle+\lambda\gamma^*-\mu\gamma]|_{\lambda=\mu=0},$$

we first remark that

$$\frac{\partial}{\partial\gamma^*}X=\exp\left[(1-\alpha)\frac{\partial}{\partial\lambda}\frac{\partial}{\partial\mu}\right]\lambda\exp[\ ]|_{\lambda=\mu=0}$$

$$=(1-\alpha)\exp\left[(1-\alpha)\frac{\partial}{\partial\lambda}\frac{\partial}{\partial\mu}\right]\frac{\partial}{\partial\mu}\exp[\ ]|$$

$$=(1-\alpha)\exp\left[(1-\alpha)\frac{\partial}{\partial\lambda}\frac{\partial}{\partial\mu}\right](-\lambda\langle n\rangle-\gamma)\exp[\ ]|,$$

from which follows

$$\frac{\partial}{\partial\gamma^*}X=-\frac{\gamma(1-\alpha)}{1+\langle n\rangle(1-\alpha)}X$$

or

$$X=X_0\exp\left[-|\gamma|^2\frac{1-\alpha}{1+\langle n\rangle(1-\alpha)}\right].$$

Here

$$X_0=\exp\left[(1-\alpha)\frac{\partial}{\partial\lambda}\frac{\partial}{\partial\mu}\right]\exp[-\lambda\mu\langle n\rangle]|_{\lambda=\mu=0}$$

$$=[1+\langle n\rangle(1-\alpha)]^{-1},$$

as one shows with a similar procedure, or by direct series expansion. Therefore,

$$\sum_0^\infty \alpha^n p(n,\vartheta,K)=\frac{1-e^{-\beta\omega}}{1-\alpha e^{-\beta\omega}}\exp\left[-|\gamma|^2\frac{1-e^{-\beta\omega}}{1-\alpha e^{-\beta\omega}}(1-\alpha)\right],$$

where

$$|\gamma|^2=\left|\int dte^{i\omega t}K(t)\right|^2,$$

and on referring to the previously used Laguerre polynomial sum formula, we obtain

$$p(n,\vartheta,K)=(1-e^{-\beta\omega})e^{-n\beta\omega}\exp[-|\gamma|^2(1-e^{-\beta\omega})]$$
$$\times L_n[-4|\gamma|^2\sinh^2(\beta\omega/2)].$$

In addition to describing the physical situation of initial thermal equilibrium, this result provides a generating function for the individual transition probabilities between oscillator energy states,

$$\sum_{n'=0}^\infty p(n,n',K)e^{-(n'-n)\beta\omega}$$
$$=\exp[-|\gamma|^2(1-e^{-\beta\omega})]L_n[-(1-e^{-\beta\omega})(e^{\beta\omega}-1)|\gamma|^2].$$

This form, and the implied transition probabilities, have already been derived in another connection,[4] and we shall only state the general result here:

$$p(n,n',K)=\frac{n_<!}{n_>!}(|\gamma|^2)^{n_>-n_<}[L_{n_<}^{(n_>-n_<)}(|\gamma|^2)]^2$$
$$\times\exp(-|\gamma|^2),$$

in which $n_>$ and $n_<$ represent the larger and smaller of the two integers n and n'.

Another kind of probability is also easily identified, that referring to the continuous spectrum of the Hermitian operator

$$q=2^{-\frac{1}{2}}(y+y^\dagger)$$

[or $p=-2^{-\frac{1}{2}}i(y-y^\dagger)$]. For this purpose, we place $\lambda=\mu=-2^{-\frac{1}{2}}p'$ and obtain

$$\langle t_2|e^{ip'q(t_1)}|t_2\rangle_\vartheta{}^K=\exp[-\tfrac{1}{2}p'^2(\langle n\rangle_\vartheta+\tfrac{1}{2})+ip'\langle q(t_1)\rangle^K],$$

with

$$\langle q(t_1)\rangle^K=2^{-\frac{1}{2}}i\left[e^{i\omega t_1}\int_{t_2}^{t_1}dte^{-i\omega t}K^*(t)\right.$$
$$\left.-e^{-i\omega t_1}\int_{t_2}^{t_1}dte^{i\omega t}K(t)\right].$$

If we multiply this result by $\exp(-ip'q')$ and integrate with respect to $p'/2\pi$ from $-\infty$ to ∞, we obtain the expectation value of $\delta[q(t_1)-q']$ which is the probability of realizing a value of $q(t_1)$ in a unit interval about q':

$$p(q't_1,\beta,K)=(\pi^{-1}\tanh\tfrac{1}{2}\beta\omega)^{\frac{1}{2}}$$
$$\times\exp[-(\tanh\tfrac{1}{2}\beta\omega)(q'-\langle q(t_1)\rangle^K)^2].$$

Still another derivation of the formula giving thermal expectation values merits attention. Now we let the return path terminate at a different time $t_2'=t_2-T$, and on regarding the resulting transformation function as a matrix, compute the trace, or rather the trace ratio

$$\mathrm{tr}\langle t_2'|t_2\rangle^{K\pm}/\mathrm{tr}\langle t_2'|t_2\rangle,$$

which reduces to unity in the absence of external forces. The action principle again describes the dependence upon $K_\pm^*(t)$, $K_\pm(t)$ through the operators $y_\pm(t)$, $y_\pm^\dagger(t)$ which are related to the forces by the solutions of the equations of motion, and, in particular,

$$y_-(t_2')=e^{-i\omega(t_2'-t_2)}y_+(t_2)-i\int_{t_2}^{t_1}dte^{i\omega(t-t_2')}K_+(t)$$
$$+i\int_{t_2'}^{t_1}dte^{i\omega(t-t_2')}K_-(t).$$

Next we recognize that the structure of the trace implies the effective boundary condition

$$y_-(t_2')=y_+(t_2).$$

[4] Julian Schwinger, Phys. Rev. **91**, 728 (1953).

Let us consider

$$\mathrm{tr}\langle t_2'|y_-(t_2')|t_2\rangle=\sum_{a'}\langle a't_2'|y_-(t_2')|a't_2\rangle,$$

where we require of the a representation only that it have no explicit time dependence. Then

$$\langle a't_2'|y_-(t_2')=\sum_{a''}\langle a'|y|a''\rangle\langle a''t_2'|$$

and

$$\mathrm{tr}\langle t_2'|y_-(t_2')|t_2\rangle=\sum_{a'a''}\langle a''t_2'|a't_2\rangle\langle a'|y|a''\rangle$$
$$=\mathrm{tr}\langle t_2'|y_+(t_2)|t_2\rangle,$$

which is the stated result.

The effective initial condition now appears as

$$y_+(t_2)=-\frac{1}{e^{-i\omega T}-1}i\left[\int_{t_2}^{t_1}dt e^{i\omega(t-t_2)}K_+(t)-\int_{t_2'}^{t_1}dt e^{i\omega(t-t_2)}K_-(t)\right],$$

and the action principle supplies the following evaluation of the trace ratio:

$$\exp\left[-i\int dt dt' K^*(t)G_0(t-t')K(t')\right]\times\exp\left[-(e^{-i\omega T}-1)^{-1}\left|\int dt e^{i\omega t}(K_+-K_-)(t)\right|^2\right],$$

where the time variable in K_+ and K_- ranges from t_2 to t_1 and from t_2' to t_1, respectively. To solve the given physical problem we require that $K_-(t)$ vanish in the interval between t_2' and t_2 so that all time integrations are extended between t_2 and t_1. Then, since

$$\langle t_2'|=\langle t_2|e^{-i\omega(t_2'-t_2)n},\quad n=y^\dagger y(t_2),$$

what has been evaluated equals

$$\mathrm{tr}\langle t_2|e^{i\omega Tn}|t_2\rangle^{K\pm}/\mathrm{tr}\langle t_2|e^{i\omega Tn}|t_2\rangle,$$

and by adding the remark that this ratio continues to exist on making the complex substitution

$$-iT\to\beta>0,$$

the desired formula emerges as

$$\mathrm{tr}\langle t_2|e^{-\beta\omega n}|t_2\rangle^{K\pm}/\mathrm{tr}\langle t_2|e^{-\beta\omega n}|t_2\rangle=\exp\left[-i\int dt dt' K^*(t)G_\vartheta(t-t')K(t')\right].$$

EXTERNAL SYSTEM

This concludes our preliminary survey of the oscillator and we turn to the specific physical problem of interest: An oscillator subjected to prescribed external forces and loosely coupled to an essentially macroscopic external system. All oscillator interactions are linear in the oscillator variables, as described by the Lagrangian operator

$$L=iy^\dagger(dy/dt)-\omega_0 y^\dagger y-y^\dagger K(t)-yK^*(t)-2^{\frac{1}{2}}qQ+L_{\mathrm{ext}},$$

in which L_{ext} characterizes the external system and $Q(t)$ is a Hermitian operator of that system.

We begin our treatment with a discussion of the transformation function $\langle t_2|t_2\rangle_{\vartheta_0\vartheta}{}^{K\pm}$ that refers initially to a thermal mixture at temperature ϑ for the external system, and to an independent thermal mixture at temperature ϑ_0 for the oscillator. The latter temperature can be interpreted literally, or as a convenient parametric device for obtaining expectation values referring to oscillator energy states. To study the effect of the coupling between the oscillator and the external system we supply the coupling term with a variable parameter λ, and compute

$$\frac{\partial}{\partial\lambda}\langle t_1|t_2\rangle^{K\pm}=-i\left\langle t_2\left|\int_{t_2}^{t_1}dt[2^{\frac{1}{2}}q_+(t)Q_+(t)-2^{\frac{1}{2}}q_-(t)Q_-(t)]\right|t_2\right\rangle^{K\pm},$$

where the distinction between the forward and return paths arises only from the application of different external forces $K_\pm(t)$ on the two segments of the closed time contour. The characterization of the external system as essentially macroscopic now enters through the assumption that this large system is only slightly affected by the coupling to the oscillator. In a corresponding first approximation, we would replace the operators $Q_\pm(t)$ by the effective numerical quantity $\langle Q(t)\rangle_\vartheta$. The phenomena that appear in this order of accuracy are comparatively trivial, however, and we shall suppose that

$$\langle Q(t)\rangle_\vartheta=0,$$

which forces us to proceed to the next approximation.

A second differentiation with respect to λ gives

$$-\frac{1}{2}\frac{\partial^2}{\partial\lambda^2}\langle t_2|t_2\rangle^{K\pm}=\left\langle t_2\left|\int_{t_2}^{t_1}dt dt'[(qQ(t)qQ(t'))_+-2q_-Q_-(t)q_+Q_+(t')+(qQ(t)qQ(t'))_-]\right|t_2\right\rangle^{K\pm}.$$

The introduction of an approximation based upon the slight disturbance of the macroscopic system converts this into

$$-\frac{1}{2}\frac{\partial^2}{\partial\lambda^2}\langle t_2|t_2\rangle^{K\pm}=\left\langle t_2\left|\int_{t_2}^{t_1}dt dt'[(y(t')y^\dagger(t))_+A_{++}(t-t')-y_-(t')y_+^\dagger(t)A_{+-}(t-t')-y_-^\dagger(t)y_+(t')A_{-+}(t-t')+(y(t')y^\dagger(t))_-A_{--}(t-t')]\right|t_2\right\rangle^{K\pm},$$

where

$$A(t-t')=\begin{pmatrix}\langle(Q(t)Q(t'))_+\rangle_\vartheta & \langle Q(t')Q(t)\rangle_\vartheta\\ \langle Q(t)Q(t')\rangle_\vartheta & \langle(Q(t)Q(t'))_-\rangle_\vartheta\end{pmatrix},$$

and we have also discarded all terms containing $y(t)y(t')$ and $y^\dagger(t)y^\dagger(t')$. The latter approximation refers to the assumed weakness of the coupling of the oscillator to the external system, for, during the many periods that are needed for the effect of the coupling to accumulate, quantities with the time dependence $e^{\pm i\omega_0(t+t')}$ will become suppressed in comparison with those varying as $e^{\pm i\omega_0(t-t')}$. At this point we ask what effective term in an action operator that refers to the closed time path of the oscillator would reproduce this approximate value of $(\partial/\partial\lambda)^2\langle t_2|t_2\rangle$ at $\lambda=0$. The complete action that satisfies this requirement, with λ^2 set equal to unity, is given by

$$W=\int_{t_2}^{t_1}dt\left[iy^\dagger\frac{dy}{dt}-\omega_0y^\dagger y-y^\dagger K-yK^*\Big|_+-\Big|_-\right]$$

$$+i\int_{t_2}^{t_1}dtdt'[(y^\dagger(t)y(t'))_+A_{++}(t-t')$$

$$-y_-^\dagger(t)y_+(t')A_{-+}(t-t')-y_-(t')y_+^\dagger(t)A_{+-}(t-t')$$

$$+(y^\dagger(t)y(t'))_-A_{--}(t-t')].$$

The application of the principle of stationary action to this action operator yields equations of motion that are nonlocal in time, namely,

$$i\frac{dy_+}{dt}-\omega_0y_++i\int_{t_2}^{t_1}dt'[A_{++}(t-t')y_+(t')$$
$$-A_{+-}(t-t')y_-(t')]=K_+(t)$$

$$i\frac{dy_-}{dt}-\omega_0y_--i\int_{t_2}^{t_1}dt'[A_{--}(t-t')y_-(t')$$
$$-A_{-+}(t-t')y_+(t')]=K_-(t),$$

together with

$$-i\frac{dy_+^\dagger}{dt}-\omega_0y_+^\dagger+i\int_{t_2}^{t_1}dt'[y_+^\dagger(t')A_{++}(t'-t)$$
$$-y_-^\dagger(t')A_{-+}(t'-t)]=K_+^*(t)$$

$$-i\frac{dy_-^\dagger}{dt}-\omega_0y_-^\dagger-i\int_{t_2}^{t_1}dt'[y_-^\dagger(t')A_{--}(t'-t)$$
$$-y_+^\dagger(t')A_{+-}(t'-t)]=K_-^*(t).$$

The latter set is also obtained by combining the formal adjoint operation with the interchange of the + and − labels attached to the operators and $K(t)$. Another significant form is conveyed by the pair of equations

$$\left(i\frac{d}{dt}-\omega_0\right)(y_--y_+)-i\int_{t_2}^{t_1}dt'(A_{--}-A_{+-})(t-t')$$
$$\times(y_--y_+)(t')=K_--K_+$$

and

$$\left(i\frac{d}{dt}-\omega_0\right)(y_++y_-)+i\int_{t_2}^{t_1}dt'(A_{++}-A_{+-})(t-t')$$
$$\times(y_++y_-)(t')-i\int_{t_2}^{t_1}dt'(A_{+-}+A_{-+})(t-t')$$
$$\times(y_--y_+)(t')=K_++K_-,$$

where

$$(A_{--}-A_{+-})(t-t')=-(A_{++}-A_{-+})(t-t')$$
$$=\langle[Q(t),Q(t')]\rangle_\vartheta\eta_-(t-t'),$$

$$(A_{++}-A_{+-})(t-t')=-(A_{--}-A_{-+})(t-t')$$
$$=\langle[Q(t),Q(t')]\rangle_\vartheta\eta_+(t-t'),$$

and

$$(A_{+-}+A_{-+})(t-t')=\langle\{Q(t),Q(t')\}\rangle_\vartheta.$$

The nonlocal character of these equations is not very marked if, for example, the correlation between $Q(t)$ and $Q(t')$ in the macroscopic system disappears when $|t-t'|$ is still small compared with the period of the oscillator. Then, since the behavior of $y(t)$ over a short time interval is given approximately by $e^{-i\omega t}$, the matrix $A(t-t')$ is effectively replaced by

$$\int_{-\infty}^{\infty}d(t-t')e^{i\omega(t-t')}A(t-t')=A(\omega),$$

and the equations of motion read

$$[i(d/dt)-\omega_-](y_--y_+)=K_--K_+,$$
$$[i(d/dt)-\omega_+](y_++y_-)-ia(y_--y_+)=K_++K_-.$$

Here we have defined

$$\omega_-=\omega_0+i(A_{--}-A_{+-})(\omega)=\omega+\tfrac{1}{2}i\gamma,$$
$$\omega_+=\omega_0-i(A_{++}-A_{+-})(\omega)=\omega-\tfrac{1}{2}i\gamma,$$

and

$$a(\omega)=(A_{+-}+A_{-+})(\omega).$$

It should be noted that $A_{+-}(\omega)$ and $A_{-+}(\omega)$ are real positive quantities since

$$A_{-+}(\omega)=\lim_{T\to\infty}\frac{1}{T}\left\langle\left(\int_{-\frac{1}{2}T}^{\frac{1}{2}T}dte^{-i\omega t}Q(t)\right)^\dagger\right.$$
$$\left.\times\left(\int_{-\frac{1}{2}T}^{\frac{1}{2}T}dte^{-i\omega t}Q(t)\right)\right\rangle$$

and

$$A_{+-}(\omega)=A_{-+}(-\omega).$$

One consequence is

$$a(\omega)=a(-\omega)\geq0.$$

It also follows from

$$\omega_--\omega_+=i(A_{--}+A_{++}-2A_{+-})(\omega)$$
$$=i(A_{-+}-A_{+-})(\omega)$$

that

$$\gamma(\omega)=A_{-+}(\omega)-A_{+-}(\omega)$$
$$=-\gamma(-\omega)$$

is real. Furthermore

$$\omega=\omega_0-\tfrac{1}{2}i(A_{++}-A_{--})(\omega),$$

where

$$(A_{++}-A_{--})(t-t')=\langle[Q(t),Q(t')]\rangle_\vartheta\epsilon(t-t')$$
$$=(A_{-+}-A_{+-})(t-t')\epsilon(t-t'),$$

so that

$$-i(A_{++}-A_{--})(\omega)=\frac{1}{\pi}P\int_{-\infty}^{\infty}\frac{d\omega'}{\omega-\omega'}\gamma(\omega'),$$

and ω emerges as the real quantity

$$\omega=\omega_0-\frac{1}{\pi}P\int_0^{\infty}\frac{\omega' d\omega'}{\omega'^2-\omega^2}\gamma(\omega').$$

We have not yet made direct reference to the nature of the expectation value for the macroscopic system, which is now taken as the thermal average:

$$\langle X\rangle_\vartheta=C\ \mathrm{tr}e^{-\beta H}X$$
$$C^{-1}=\mathrm{tr}e^{-\beta H},$$

where H is the energy operator of the external system. The implication for the structure of the expectation values is contained in

$$\langle Q(t)Q(t')\rangle_\vartheta=C\ \mathrm{tr}e^{-\beta H}Q(t)Q(t')$$
$$=\langle Q(t')Q(t+i\beta)\rangle_\vartheta,$$

which employs the formal property

$$e^{-\beta H}Q(t)e^{\beta H}=Q(t+i\beta).$$

On introducing the time Fourier transforms, however, this becomes the explicit relation

$$A_{-+}(\omega)=e^{\beta\omega}A_{+-}(\omega),$$

and we conclude that

$$e^{-\frac{1}{2}\beta\omega}A_{-+}(\omega)=e^{\frac{1}{2}\beta\omega}A_{+-}(\omega)$$
$$=a(\omega)/2\cosh\tfrac{1}{2}\beta\omega,$$

which is a positive even function of ω. As a consequence, we have

$$\gamma(\omega)=a(\omega)\tanh\tfrac{1}{2}\beta\omega,$$
$$\geq 0,\quad \beta\omega>0,$$

which can also be written as

$$a(\omega)=2\gamma(\omega)[(e^{\beta\omega}-1)^{-1}+\tfrac{1}{2}].$$

The net result of this part of the discussion is to remove all explicit reference to the external system as a dynamical entity. We are given effective equations of motion for y_+ and y_- that contain the prescribed external forces and three parameters, the angular frequency $\omega(\simeq\omega_0)$, γ, and a, the latter pair being related by the temperature of the macroscopic system. The accompanying boundary conditions are

$$(y_--y_+)(t_1)=0$$

and, for the choice of an initial thermal mixture,

$$y_-(t_2)=e^{\beta_0\omega}y_+(t_2).$$

We now find that

$$(y_--y_+)(t)=i\int_{t_2}^{t_1}dt'e^{i\omega_-(t'-t)}\eta_-(t-t')(K_--K_+)(t'),$$

which supplies the initial condition for the second equation of motion,

$$(y_++y_-)(t_2)=\coth(\tfrac{1}{2}\beta_0\omega)i\int_{t_2}^{t_1}dte^{i\omega_-(t-t_2)}(K_--K_+)(t),$$

and the required solution is given by

$$i(y_++y_-)(t)$$
$$=\int_{t_2}^{t_1}dt'e^{-i\omega_+(t-t')}\eta_+(t-t')(K_++K_-)(t')$$
$$-\coth(\tfrac{1}{2}\beta\omega)\int_{t_2}^{t_1}dt'[e^{-i\omega_+(t-t')}\eta_+(t-t')$$
$$+e^{i\omega_-(t'-t)}\eta_-(t-t')](K_--K_+)(t')$$
$$+(\coth\tfrac{1}{2}\beta\omega-\coth\tfrac{1}{2}\beta_0\omega)\int_{t_2}^{t_1}dt'e^{-i\omega_+(t-t_2)}$$
$$\times e^{i\omega_-(t'-t_2)}(K_--K_+)(t').$$

The corresponding solutions for $y_\pm^\dagger(t)$ are obtained by interchanging the $\pm$ labels in the formal adjoint equation.

The differential dependence of the transformation function $\langle t_2|t_2\rangle_{\vartheta_0\vartheta}{}^{K\pm}$ upon the external forces is described by these results, and the explicit formula obtained on integration is

$$\langle t_2|t_2\rangle_{\vartheta_0\vartheta}{}^{K\pm}$$
$$=\exp\left[-i\int dtdt'K^*(t)G_{\vartheta_0\vartheta}(t-t_2,t'-t_2)K(t')\right],$$

where $[n_0=\langle n\rangle_{\vartheta_0},\ n=\langle n\rangle_\vartheta]$

$$iG_{\vartheta_0\vartheta}(t-t_2,t'-t_2)$$
$$=e^{-i\omega_+(t-t')}\eta_+(t-t')\begin{pmatrix}n+1, & -n\\ -n-1, & n\end{pmatrix}$$
$$+e^{-i\omega_-(t-t')}\eta_-(t-t')\begin{pmatrix}n, & -n\\ -n-1, & n+1\end{pmatrix}$$
$$+e^{-i\omega_+(t-t_2)}e^{i\omega_-(t'-t_2)}(n_0-n)\begin{pmatrix}1 & -1\\ -1 & 1\end{pmatrix}.$$

Another way of presenting this result is

$$iG_{\vartheta_0\vartheta}(t-t_2,t'-t_2)$$
$$=e^{-i\omega(t-t')}e^{-\frac{1}{2}\gamma|t-t'|}\begin{pmatrix}\eta_+(t-t')+n, & -n\\ -n-1, & \eta_-(t-t')+n\end{pmatrix}$$
$$+e^{-i\omega(t-t')}e^{-\gamma[\frac{1}{2}(t+t')-t_2]}(n_0-n)\begin{pmatrix}1 & -1\\ -1 & 1\end{pmatrix},$$

although the simplest description of G is supplied by

the differential equation

$$\left[\left(i\frac{d}{dt}-\omega_+\right)G-\delta(t-t')\begin{pmatrix}1 & 0\\ 0 & -1\end{pmatrix}\right]\left(-i\frac{d^T}{dt'}-\omega_-\right)$$
$$=-i\delta(t-t')\gamma\begin{pmatrix}n, & -n\\ -n-1, & n+1\end{pmatrix},$$

(where d^T indicates differentiation to the left) in conjunction with the initial value

$$iG_{\vartheta_0\vartheta}(0,0)=\begin{pmatrix}n_0+\frac{1}{2}, & -n_0\\ -n_0-1, & n_0+\frac{1}{2}\end{pmatrix}$$

and the boundary conditions

$$[i(d/dt)-\omega_+]G=0, \quad t>t'$$
$$[i(d/dt)-\omega_-]G=0, \quad t<t'.$$

A more symmetrical version of this differential equation is given by

$$\left(i\frac{d}{dt}-\omega_+\right)\left(-i\frac{d}{dt'}-\omega_-\right)G$$
$$-\begin{pmatrix}1 & 0\\ 0 & -1\end{pmatrix}\left(i\frac{d}{dt}-\omega\right)\delta(t-t')$$
$$=-i\delta(t-t')\gamma\begin{pmatrix}n+\frac{1}{2}, & -n\\ -n-1, & n+\frac{1}{2}\end{pmatrix}.$$

We note the vanishing sum of all G elements, and that the role of complex conjugation in exchanging the two segments of the closed time path is expressed by

$$-\begin{pmatrix}0 & 1\\ 1 & 0\end{pmatrix}G(t',t)^{T*}\begin{pmatrix}0 & 1\\ 1 & 0\end{pmatrix}=G(t,t'),$$

which is to say that

$$-G(t',t)_{+-}{}^*=G(t,t')_{+-}, \quad -G(t',t)_{-+}{}^*=G(t,t')_{-+}$$
$$-G(t',t)_{--}{}^*=G(t,t')_{++}.$$

It will be observed that only when

$$\langle n\rangle_{\vartheta_0}=\langle n\rangle_\vartheta$$

is $G_{\vartheta_0\vartheta}(t-t_2,\,t'-t_2)$ independent of t_2 and a function of $t-t'$. This clearly refers to the initial physical situation of thermal equilibrium between the oscillator and the external system at the common temperature $\vartheta_0=\vartheta>0$, which equilibrium persists in the absence of external forces. If the initial circumstances do not constitute thermal equilibrium, that will be established in the course of time at the macroscopic temperature $\vartheta>0$. Thus, all reference to the initial oscillator temperature disappears from $G_{\vartheta_0\vartheta}(t-t_2,\,t'-t_2)$ when, for fixed $t-t'$,

$$\gamma[\tfrac{1}{2}(t+t')-t_2]\gg 1.$$

The thermal relaxation of the oscillator energy is derived from

$$\langle t_2|y^\dagger y(t_1)|t_2\rangle_{\vartheta_0\vartheta}=\frac{\delta}{\delta K_-(t_1)}\frac{\delta}{\delta K_+{}^*(t_1)}\langle t_2|t_2\rangle_{\vartheta_0\vartheta}{}^{K\pm}\Big|_{K_\pm=0}$$
$$=-iG_{\vartheta_0\vartheta}(t_1-t_2,\,t_1-t_2)_{+-},$$

and is expressed by

$$\langle n(t_1)\rangle=\langle n\rangle_\vartheta+(\langle n\rangle_{\vartheta_0}-\langle n\rangle_\vartheta)e^{-\gamma(t_1-t_2)}.$$

The previously employed technique of impulsive forces applied at the time t_1 gives the more general result

$$\langle t_2|\exp[-i(\lambda y^\dagger(t_1)+\mu y(t_1))]|t_2\rangle_{\vartheta_0\vartheta}{}^K$$
$$=\exp\left[-\lambda\mu(\langle n(t_1)\rangle+\tfrac{1}{2})+\lambda\int_{t_2}^{t_1}dte^{i\omega_-(t_1-t)}K^*(t)\right.$$
$$\left.-\mu\int_{t_2}^{t_1}dte^{-i\omega_+(t_1-t)}K(t)\right],$$

from which a variety of probability distributions and expectation values can be obtained.

The latter calculation illustrates a general characteristic of the matrix $G(t,t')$, which is implied by the lack of dependence on the time t_1. Indeed, such a terminal time need not appear explicitly in the structure of the transformation function $\langle t_2|t_2\rangle^{K\pm}$ and all time integrations can range from t_2 to $+\infty$. Then t_1 is implicit as the time beyond which K_+ and K_- are identified, and the structure of G must be such as to remove any reference to a time greater than t_1. In the present situation, the use of an impulsive force at t_1 produces, for example, the term

$$\int_{t_2}^{\infty}dtG(t_1-t_2,\,t-t_2)K(t),$$

in which K_+ and K_- are set equal. Hence it is necessary that

$$G(t,t')\begin{pmatrix}1\\ 1\end{pmatrix}=0, \quad t<t'$$

and similarly that

$$\overbrace{1\quad 1}\,G(t,t')=0, \quad t>t',$$

which says that adding the columns of $G(t,t')$ gives retarded functions of $t-t'$, while the sum of rows supplies a vector that is an advanced function of $t-t'$. In each instance, the two components must have a zero sum. These statements are immediately verified for the explicitly calculated $G_{\vartheta_0\vartheta}(t-t_2,\,t'-t_2)$ and follow more generally from the operator construction

$$iG=\begin{pmatrix}\langle(y(t)y^\dagger(t'))_+\rangle, & -\langle y^\dagger(t')y(t)\rangle\\ -\langle y(t)y^\dagger(t')\rangle, & \langle(y(t)y^\dagger(t'))_-\rangle\end{pmatrix},$$

for, as we have already noted in connection with Q products,

$$(y(t)y^\dagger(t'))_+-y^\dagger(t')y(t)=y(t)y^\dagger(t')-(y(t)y^\dagger(t'))_-$$
$$=\eta_+(t-t')[y(t),y^\dagger(t')]$$

and

$$(y(t)y^\dagger(t'))_+ - y(t)y^\dagger(t') = y^\dagger(t')y(t) - (y(t)y^\dagger(t'))_- = -\eta_-(t-t')[y(t),y^\dagger(t')].$$

Our results show, incidentally, that

$$\langle[y(t),y^\dagger(t')]\rangle_{\vartheta_0\vartheta} = e^{-i\omega_+(t-t')}\eta_+(t-t') + e^{-i\omega_-(t-t')}\eta_-(t-t') = e^{-i\omega(t-t')}e^{-\frac{1}{2}\gamma|t-t'|}.$$

Another general property can be illustrated by our calculation, the positiveness of $-iG(t,t')_{+-}$,

$$-\int dtdt' K(t) iG(t-t_2, t'-t_2)_{+-}K^*(t') = \left\langle t_2\left|\left(\int dtK(t)y(t)\right)^\dagger\left(\int dtK(t)y(t)\right)\right|t_2\right\rangle > 0.$$

We have found that

$$-iG_{\vartheta_0\vartheta}(t-t_2, t'-t_2)_{+-} = \exp\{-i\omega(t-t') - \gamma[\tfrac{1}{2}(t+t')-t_2]\}\langle n\rangle_\vartheta + e^{-i\omega(t-t')}[e^{-\frac{1}{2}\gamma|t-t'|} - e^{-\gamma[\frac{1}{2}(t+t')-t_2]}]\langle n\rangle_{\vartheta_0},$$

and it is clearly necessary that each term obey separately the positiveness requirement. The first term is trivial,

$$\int dtdt' K(t)\exp\{-i\omega(t-t') - \gamma[\tfrac{1}{2}(t+t')-t_2]\}K^*(t') = \left|\int dte^{-i\omega_+(t-t_2)}K(t)\right|^2 > 0,$$

and the required property of the second term follows from the formula

$$e^{-\frac{1}{2}\gamma|t-t'|} - e^{-\gamma[\frac{1}{2}(t+t')-t_2]} = \frac{2\gamma}{\pi}\int_0^\infty d\omega' \frac{\sin\omega'(t-t_2)\sin\omega'(t'-t_2)}{\omega'^2+(\frac{1}{2}\gamma)^2}.$$

All the information that has been obtained about the oscillator is displayed on considering the forces

$$K_\pm(t) = \lambda_\pm(t) + K(t), \quad K_\pm^*(t) = \mu_\pm(t) + K^*(t),$$

and making explicit the effects of $\lambda_\pm(t)$, $\mu_\pm(t)$ by equivalent time-ordered operators:

$$\left\langle t_2\left|\left(\exp\left[i\int_{t_2}^\infty dt(\lambda_- y^\dagger + \mu_- y)\right]\right)_- \times\left(\exp\left[-i\int_{t_2}^\infty dt(\lambda_+ y^\dagger + \mu_+ y)\right]\right)_+\right|t_2\right\rangle^K_{\vartheta_0\vartheta}$$
$$= \exp\left[-i\int dtdt' \mu(t) G(t-t_2, t'-t_2)\lambda(t') + \int dtdt' K^*(t)e^{-i\omega_-(t-t')}\eta_-(t-t')(\lambda_+-\lambda_-)(t') - \int dtdt'(\mu_+-\mu_-)(t)e^{-i\omega_+(t-t')}\eta_+(t-t')K(t')\right].$$

This is a formula for the direct computation of expectation values of general functions of $y(t)$ and $y^\dagger(t)$. A less explicit but simpler result can also be given by means of expectation values for functions of the operators

$$[i(d/dt)-\omega_+]y(t) - K(t) = K_f(t),$$
$$[-i(d/dt)-\omega_-]y^\dagger(t) - K^*(t) = K_f^\dagger(t).$$

Let us recognize at once that

$$\langle K_f(t)\rangle = 0, \quad \langle K_f^\dagger(t)\rangle = 0,$$

and therefore that the fluctuations of $y(t)$, $y^\dagger(t)$ can be ascribed to the effect of the forces K_f, $K_f^\dagger$, which appear as the quantum analogs of the random forces in the classical Langevin approach to the theory of the Brownian motion. The change in viewpoint is accomplished by introducing

$$\lambda_\pm(t) = [i(d/dt)-\omega_-]u_\pm(t)$$
$$\mu_\pm(t) = [-i(d/dt)-\omega_+]v_\pm(t),$$

where we assume, just for simplicity, that the functions $u(t)$, $v(t)$ vanish at the time boundaries. Then, partial time integrations will replace the operators y, $y^\dagger$ with K_f, $K_f^\dagger$.

To carry this out, however, we need the following lemma on time-ordered products:

$$\left(\exp\left[\int dt(A(t)+(d/dt)B(t))\right]\right)_+ = \left(\exp\left[\int dtA(t)\right]\right)_+ \exp\left(\int dt[A+\tfrac{1}{2}(dB/dt), B]\right),$$

which involves the unessential assumption that $B(t)$ vanishes at the time terminals, and the hypothesis that $[A(t), B(t)]$ and $[dB(t)/dt, B(t)]$ are commutative with all the other operators. The proof is obtained by replacing $B(t)$ with $\lambda B(t)$ and differentiating with respect to λ,

$$\frac{\partial}{\partial\lambda}\left(\exp\left[\int_{t_2}^{t_1} dt\left(A+\lambda\frac{d}{dt}B\right)\right]\right)_+ = \int_{t_2}^{t_1} dt\left(\exp\left[\int_t^{t_1}\right]\right)_+ \frac{d}{dt}B(t)\left(\exp\left[\int_{t_2}^{t}\right]\right)_+.$$

Then, a partial integration yields

$$\int_{t_2}^{t_1} dt\left(\exp\left[\int_t^{t_1}\right]\right)_+\left[A(t)+\lambda\frac{dB(t)}{dt}, B(t)\right] \times\left(\exp\left[\int_{t_2}^{t}\right]\right)_+ = \left(\exp\left[\int_{t_2}^{t_1}\right]\right)_+\int_{t_2}^{t_1} dt\left[A+\lambda\frac{dB}{dt}, B\right],$$

according to the hypothesis, and the stated result follows on integrating this differential equation.

The structure of the lemma is given by the rearrangement

$$-i(\lambda y^\dagger+\mu y)=-i[u(K^*+K_f^\dagger)+v(K+K_f)] \\ +(d/dt)(uy^\dagger-vy),$$

and we immediately find a commutator that is a multiple of the unit operator,

$$[A+(d/dt)B, B]=-i[\lambda y^\dagger+\mu y, uy^\dagger-vy] \\ =-i(\mu u+\lambda v)\rightarrow -2iv(i(d/dt)-\omega)u.$$

The last form involves discarding a total time derivative that will not contribute to the final result. To evaluate $[A,B]$ we must refer to the meaning of K_f and $K_f^\dagger$ that is supplied by the actual equations of motion,

$$K_f(t)=Q(t)+(\omega_0-\omega_+)y(t)$$
$$K_f^\dagger(t)=Q(t)+(\omega_0-\omega_-)y^\dagger(t),$$

for then

$$[A(t),B(t)]=-i[u(\omega_0-\omega_-)y^\dagger+v(\omega_0-\omega_+)y,\ uy^\dagger-vy] \\ =2ivu(\omega-\omega_0),$$

which is also proportional to the unit operator. Accordingly,

$$\left(\exp\left[-i\int dt(\lambda y^\dagger+\mu y)\right]\right)_+$$
$$=\left(\exp\left[-i\int dt[u(K^*+K_f^\dagger)+v(K+K_f)]\right]\right)_+$$
$$\times\exp\left[i(\omega-\omega_0)\int dtvu-i\int dtv\left(i\frac{d}{dt}-\omega\right)u\right],$$

and complex conjugation yields the analogous result for negatively time-ordered products.

With the aid of the differential equation obeyed by G, we now get

$$\left\langle t_2\left|\left(\exp\left[i\int dt(uK_f^\dagger+vK_f)\right]\right)_-\right.\right.$$
$$\left.\left.\times\left(\exp\left[-i\int dt(uK_f^\dagger+vK_f)\right]\right)_+\right|t_2\right\rangle^K_{\vartheta_0\vartheta}$$
$$=\exp\left[-i\int_{t_2}^{t_1} dtv(t)\kappa u(t)\right],$$

where

$$\kappa=\gamma\begin{pmatrix} n+\frac{1}{2}, & -n \\ -n-1, & n+\frac{1}{2}\end{pmatrix}+i(\omega-\omega_0)\begin{pmatrix}1 & 0\\ 0 & -1\end{pmatrix}.$$

The elements of this matrix are also expressed by

$$\kappa\delta(t-t')=\begin{pmatrix}\langle(K_f(t)K_f^\dagger(t'))_+\rangle, & -\langle K_f^\dagger(t')K_f(t)\rangle \\ -\langle K_f(t)K_f^\dagger(t')\rangle, & \langle(K_f(t)K_f^\dagger(t'))_-\rangle\end{pmatrix}.$$

Such expectation values are to be understood as effective evaluations that serve to describe the properties of the oscillator under the circumstances that validate the various approximations that have been used.

It will be observed that when n is sufficiently large to permit the neglect of all other terms,

$$\kappa\simeq\tfrac{1}{2}a\begin{pmatrix}1 & -1\\ -1 & 1\end{pmatrix}\quad[\tfrac{1}{2}a=\gamma(n+\tfrac{1}{2})],$$

and the sense of operator multiplication is no longer significant. This is the classical limit, for which

$$\left\langle\exp\left[-i\int dt(uK_f^\dagger+vK_f)\right]\right\rangle_\vartheta \\ =\exp\left[-\int dt\tfrac{1}{2}av(t)u(t)\right],$$

where we have placed $u_+-u_-=u$, $v_+-v_-=v$. On introducing real components of the random force

$$K_f=2^{-\frac{1}{2}}(K_1+iK_2),\quad K_f^\dagger=2^{-\frac{1}{2}}(K_1-iK_2),$$

the classical limiting result reads

$$\left\langle\exp\left[-i\int dt(u_1K_1+u_2K_2)\right]\right\rangle_\vartheta \\ =\exp\left[-\int dt\tfrac{1}{4}a(u_1^2+u_2^2)\right].$$

The fluctuations at different times are independent. If we consider time-averaged forces,

$$\bar{K}=\frac{1}{\Delta t}\int_t^{t+\Delta t}dt'K(t'),$$

we find by Fourier transformation that

$$\langle\delta(\bar{K}_1-K_1')\delta(\bar{K}_2-K_2')\rangle_\vartheta=\frac{\Delta t}{\pi a}\exp\left[-\frac{\Delta t}{a}(K_1'^2+K_2'^2)\right],$$

which is the Gaussian distribution giving the probability that the force averaged over a time interval Δt will have a value within a small neighborhood of the point K'. In this classical limit the fluctuation constant a is related to the damping or dissipation constant γ and the macroscopic temperature ϑ by

$$a=(2\gamma/\omega)\vartheta.$$

Our simplified equations can also be applied to situations in which the external system is not at thermal equilibrium. To see this possibility let us return to the real positive functions $A_{-+}(\omega)$, $A_{+-}(\omega)$ that describe the external system and remark that, generally,

$$\frac{A_{-+}(-\omega)}{A_{+-}(-\omega)}=\left[\frac{A_{-+}(\omega)}{A_{+-}(\omega)}\right]^{-1}\geq 0.$$

These properties can be expressed by writing

$$A_{-+}(\omega)/A_{+-}(\omega)=e^{\omega\beta(\omega)},$$

where $\beta(\omega)$ is a real even function that can range from $-\infty$ to $+\infty$. When only one value of ω is of interest, all conceivable situations for the external system can be described by the single parameter β, the reciprocal of which appears as an effective temperature of the macroscopic system. A new physical domain that appears in this way is characterized by negative temperature, $\beta<0$. Since a is an intrinsically positive constant, it is γ that will reverse sign

$$-\gamma=a(1-e^{-|\beta|\omega})/(1+e^{-|\beta|\omega})>0,$$

and the effect of the external system on the oscillator changes from damping to amplification.

We shall discuss the following physical sequence. At time t_2 the oscillator, in a thermal mixture of states at temperature ϑ_0, is acted on by external forces which are present for a time, short in comparison with $1/|\gamma|$. After a sufficiently extended interval $\sim(t_1-t_2)$ such that the amplification factor or gain is very large,

$$k=e^{\frac{1}{2}|\gamma|(t_1-t_2)}\gg 1,$$

measurements are made in the neighborhood of time t_1. A prediction of all such measurements is contained in the general expectation value formula. Approximations that convey the physical situation under consideration are given by

$$\int dtdt'(\mu_+-\mu_-)(t)e^{-i\omega_+(t-t')}\eta_+(t-t')K(t')$$

$$\simeq k\int dt(\mu_+-\mu_-)(t)e^{-i\omega t}\int dt'e^{i\omega t'}K(t'),$$

$$\int dtdt'K^*(t)e^{-i\omega_-(t-t')}\eta_-(t-t')(\lambda_+-\lambda_-)(t')$$

$$\simeq k\int dtK^*(t)e^{-i\omega t}\int dt'e^{i\omega t'}(\lambda_+-\lambda_-)(t'),$$

and

$$i\int dtdt'\mu(t)G(t-t_2,t'-t_2)\lambda(t')$$

$$\simeq k^2(\langle n\rangle_{\vartheta_0}+(1-e^{-|\beta|\omega})^{-1})\int dt(\mu_+-\mu_-)(t)e^{-i\omega t}$$

$$\times\int dt'(\lambda_+-\lambda_-)(t')e^{i\omega t'}.$$

From the appearance of the combinations $\mu_+-\mu_-=\mu$, $\lambda_+-\lambda_-=\lambda$ only, we recognize that noncommutativity of operator multiplication is no longer significant, and thus the motion of the oscillator has been amplified to the classical level. To express the consequences most simply, we write

$$y(t)=ke^{-i\omega t}(y_s+y_n)$$
$$y^\dagger(t)=ke^{i\omega t}(y_s^*+y_n^*),$$

with

$$y_s=-i\int_{t_2}^{\infty}dt'e^{i\omega t'}K(t'),$$

and, on defining

$$u=k\int dte^{i\omega t}\lambda(t),\quad v=k\int dte^{-i\omega t}\mu(t),$$

we obtain the time-independent result

$$\langle\exp[-i(uy_n^*+vy_n)]\rangle=\exp[-(\langle n\rangle_{\vartheta_0}+(1-e^{-|\beta|\omega})^{-1})vu],$$

which implies that

$$\langle y_n\rangle=\langle y_n^*\rangle=0$$
$$\langle|y_n|^2\rangle=\langle n\rangle_{\vartheta_0}+(1-e^{-|\beta|\omega})^{-1}\geq\langle n\rangle_{\vartheta_0}+1.$$

Thus, the oscillator coordinate $y(t)$ is the amplified superposition of two harmonic terms, one of definite amplitude and phase (signal), the other with random amplitude and phase (noise), governed by a two-dimensional Gaussian probability distribution.

These considerations with regard to amplification can be viewed as a primitive model of a maser device,[5] with the oscillator corresponding to a single mode of a resonant electromagnetic cavity, and the external system to an atomic ensemble wherein, for a selected pair of levels, the thermal population inequality is reversed by some means such as physical separation or electromagnetic pumping.

AN IMPROVED TREATMENT

In this section we seek to remove some of the limitations of the preceding discussion. To aid in dealing successfully with the nonlocal time behavior of the oscillator, it is convenient to replace the non-Hermitian operator description with one employing Hermitian operators. Accordingly, we begin the development again, now using the Lagrangian operator

$$L=p(dq/dt)-\tfrac{1}{2}(p^2+\omega_0^2q^2)+qF(t)+qQ+L_{\text{ext}},$$

where Q has altered its meaning by a constant factor. One could also include an external prescribed force that is coupled to p. We repeat the previous approximate construction of the transformation function $\langle t_2|t_2\rangle_{\vartheta_0}{}^{F\pm}$ which proceeds by the introduction of an effective action operator that retains only the simplest correlation aspects of the external system, as comprised in

$$A(t-t')=\begin{pmatrix}\langle(Q(t)Q(t'))_+\rangle_\vartheta & \langle Q(t')Q(t)\rangle_\vartheta\\ \langle Q(t)Q(t')\rangle_\vartheta & \langle(Q(t)Q(t'))_-\rangle_\vartheta\end{pmatrix}.$$

[5] A similar model has been discussed recently by R. Serber and C. H. Townes, *Symposium on Quantum Electronics* (Columbia University Press, New York, 1960).

The action operator, with no other approximations, is

$$W=\int_{t_2}^{t_1} dt\left[p\frac{dq}{dt}-\tfrac{1}{2}(p^2+\omega_0^2q^2)+qF(t)\,|_+-|_-\right]$$

$$+\tfrac{1}{2}i\int_{t_2}^{t_1} dt dt'[(q(t)q(t'))_+A_{++}(t-t')$$

$$-2q_-(t)q_+(t')A_{-+}(t-t')+(q(t)q(t'))_-A_{--}(t-t')],$$

and the implied equations of motion, presented as second-order differential equations after eliminating

$$p=dq/dt,.$$

are

$$\left(\frac{d^2}{dt^2}+\omega_0^2\right)q_+(t)$$

$$-i\int_{t_2}^{t_1} dt'[A_{++}(t-t')q_+(t')-A_{+-}(t-t')q_-(t')]=F_+(t)$$

and

$$\left(\frac{d^2}{dt'}+\omega_0^2\right)q_-(t)$$

$$+i\int_{t_2}^{t_1} dt[A_{--}(t-t')q_-(t')-A_{-+}(t-t')q_+(t')]=F_-(t).$$

It will be seen that the adjoint operation is equivalent to the interchange of the $\pm$labels.

We define

$$-iA_r(t-t')=\langle[Q(t),Q(t')]\rangle_\beta\eta_+(t-t')$$
$$=A_{++}-A_{+-}=A_{-+}-A_{--}$$

and

$$-iA_a(t-t')=-\langle[Q(t),Q(t')]\rangle_\beta\eta_-(t-t')$$
$$=A_{++}-A_{-+}=A_{+-}-A_{--},$$

together with

$$a(t-t')=\langle\{Q(t),Q(t')\}\rangle_\beta$$
$$=A_{+-}+A_{-+},$$

which enables us to present the integro-differential equations as

$$\left(\frac{d^2}{dt^2}+\omega_0^2\right)(q_--q_+)(t)-\int_{t_2}^{t_1} dt'A_a(t-t')(q_--q_+)(t')$$
$$=(F_--F_+)(t)$$

and

$$\left(\frac{d^2}{dt^2}+\omega_0^2\right)(q_++q_-)(t)-\int_{t_2}^{t_1} dt'A_r(t-t')(q_++q_-)(t')$$

$$+i\int_{t_2}^{t_1} dt'a(t-t')(q_--q_+)(t')=(F_++F_-)(t).$$

The accompanying boundary conditions are

$$(q_--q_+)(t_1)=0,\quad (d/dt)(q_--q_+)(t_1)=0$$

and

$$q_-(t_2)=q_+(t_2)\cosh\beta_0\omega_0+\frac{i}{\omega_0}\frac{d}{dt}q_+(t_2)\sinh\beta_0\omega_0$$

$$\frac{d}{dt}q_-(t_2)=-i\omega_0q_+(t_2)\sinh\beta_0\omega_0+\frac{d}{dt}q_+(t_2)\cosh\beta_0\omega_0,$$

or, more conveniently expressed,

$$(q_++q_-)(t_2)=\frac{i}{\omega_0}\coth(\tfrac{1}{2}\beta_0\omega_0)\frac{d}{dt}(q_--q_+)(t_2)$$

$$\frac{d}{dt}(q_++q_-)(t_2)=-i\omega_0\coth(\tfrac{1}{2}\beta_0\omega_0)(q_--q_+)(t_2),$$

which replace the non-Hermitian relations

$$y_-(t_2)=e^{\beta_0\omega_0}y_+(t_2),\quad y_-^\dagger(t_2)=e^{-\beta_0\omega_0}y_+^\dagger(t_2).$$

Note that it is the intrinsic oscillator frequency ω_0 that appears here since the initial condition refers to a thermal mixture of unperturbed oscillator states.

The required solution of the equation for q_--q_+ can be written as

$$(q_--q_+)(t)=\int_{-\infty}^{\infty} dt'G_a(t-t')(F_--F_+)(t'),$$

where $G_a(t-t')$ is the real Green's function defined by

$$\left(\frac{d^2}{dt^2}+\omega_0^2\right)G_a(t-t')-\int_{-\infty}^{\infty} d\tau A_a(t-\tau)G_a(\tau-t')=\delta(t-t')$$

and

$$G_a(t-t')=0,\quad t>t'.$$

Implicit is the time t_1 as one beyond which F_--F_+ equals zero. The initial conditions for the second equation, which this solution supplies, are

$$(q_++q_-)(t_2)=\frac{i}{\omega_0}\coth(\tfrac{1}{2}\beta_0\omega_0)\int_{t_2}^{\infty} dt'\frac{\partial}{\partial t_2}$$
$$\times G_a(t_2-t')(F_--F_+)(t')$$

and

$$\frac{d}{dt}(q_++q_-)(t_2)$$
$$=-i\omega_0\coth(\tfrac{1}{2}\beta_0\omega_0)\int_{t_2}^{\infty} dt'G_a(t_2-t')(F_--F_+)(t').$$

The Green's function that is appropriate for the equation obeyed by q_++q_- is defined by

$$\left(\frac{d^2}{dt^2}+\omega_0^2\right)G_r(t-t')-\int_{-\infty}^{\infty} d\tau A_r(t-\tau)G_r(\tau-t')=\delta(t-t'),$$

$$G_r(t-t')=0,\quad t<t',$$

and the two real functions are related by

$$G_a(t-t')=G_r(t'-t).$$

The desired solution of the second differential equation is

$$(q_++q_-)(t)=\int_{t_2}^{\infty}dt'G_r(t-t')(F_++F_-)(t')$$
$$-i\int_{t_2}^{\infty}dt'w(t-t_2,\,t'-t_2)(F_--F_+)(t'),$$

where

$$w(t-t_2,\,t'-t_2)$$
$$=\int_{t_2}^{\infty}d\tau d\tau' G_r(t-\tau)a(\tau-\tau')G_a(\tau'-t')$$
$$+\frac{1}{\omega_0}\coth(\tfrac{1}{2}\beta_0\omega_0)\left[\frac{\partial}{\partial t_2}G_r(t-t_2)\frac{\partial}{\partial t_2}G_a(t_2-t')+\omega_0^2G_r(t-t_2)G_a(t_2-t')\right]$$

is a real symmetrical function of its two arguments.

The differential description of the transformation function that these solutions imply is indicated by

$$\delta_{F_\pm}\langle t_2|t_2\rangle^{F\pm}=i\left\langle t_2\left|\int dt(\delta F_+q_+-\delta F_-q_-)\right|t_2\right\rangle$$
$$=-\tfrac{1}{2}i\left\langle t_2\left|\int dt[\delta(F_--F_+)(q_++q_-)+\delta(F_++F_-)(q_--q_+)]\right|t_2\right\rangle,$$

and the result of integration is

$$\langle t_2|t_2\rangle_{\vartheta_0\vartheta}{}^{F\pm}$$
$$=\exp\left\{-\tfrac{1}{2}i\int dtdt'(F_--F_+)(t)G_r(t-t')(F_++F_-)(t')\right.$$
$$\left.-\tfrac{1}{4}\int dtdt'(F_--F_+)(t)w(t-t_2,\,t'-t_2)(F_--F_+)(t')\right\}.$$

This can also be displayed in the matrix form

$$\langle t_2|t_2\rangle_{\vartheta_0\vartheta}{}^{F\pm}=\exp\left\{\tfrac{1}{2}i\int dtdt'F(t)G_{\vartheta_0\vartheta}(t-t_2,\,t'-t_2)F(t')\right\},$$

with

$$G_{\vartheta_0\vartheta}(t-t_2,\,t'-t_2)$$
$$=\tfrac{1}{2}G_r(t-t')\begin{pmatrix}1&1\\-1&-1\end{pmatrix}+\tfrac{1}{2}G_a(t-t')\begin{pmatrix}1&-1\\1&-1\end{pmatrix}$$
$$+\tfrac{1}{2}iw(t-t_2,\,t'-t_2)\begin{pmatrix}1&-1\\-1&1\end{pmatrix}.$$

The latter obeys

$$G(t',t)^T=G(t,t')$$
$$-\begin{pmatrix}0&1\\1&0\end{pmatrix}G(t,t')^*\begin{pmatrix}0&1\\1&0\end{pmatrix}=G(t,t'),$$

and its elements are given by

$$G=i\begin{pmatrix}\langle(q(t)q(t'))_+\rangle_{\vartheta_0\vartheta}, & -\langle q(t')q(t)\rangle_{\vartheta_0\vartheta}\\ -\langle q(t)q(t')\rangle_{\vartheta_0\vartheta}, & \langle(q(t)q(t'))_-\rangle_{\vartheta_0\vartheta}\end{pmatrix}.$$

We note the identifications

$$G_r(t-t')=i\langle[q(t),q(t')]\rangle\eta_+(t-t')$$
$$G_a(t-t')=-i\langle[q(t),q(t')]\rangle\eta_-(t-t')$$
$$w(t-t_2,\,t'-t_2)=\langle\{q(t),q(t')\}\rangle.$$

It is also seen that the sum of the columns of G is proportional to $G_r(t-t')$, while the sum of the rows contains only $G_a(t-t')$.

We shall suppose that $G_r(t-t')$ can have no more than exponential growth, $\sim e^{\alpha(t-t')}$, as $t-t'\to\infty$. Then the complex Fourier transform

$$G(\zeta)=\int_{-\infty}^{\infty}d(t-t')e^{i\zeta(t-t')}G_r(t-t')$$

exists in the upper half-plane

$$\mathrm{Im}\zeta>\alpha$$

and is given explicitly by

$$G(\zeta)=[\omega_0^2-\zeta^2-A(\zeta)]^{-1}.$$

Here

$$A(\zeta)=\int_{-\infty}^{\infty}d(t-t')e^{i\zeta(t-t')}A_r(t-t')$$
$$=i\int_0^{\infty}d\tau e^{i\zeta\tau}\int_{-\infty}^{\infty}\frac{d\omega}{2\pi}e^{-i\omega\tau}(A_{-+}-A_{+-})(\omega)$$
$$=\int_{-\infty}^{\infty}\frac{d\omega}{2\pi}\frac{(A_{-+}-A_{+-})(\omega)}{\omega-\zeta}$$

or, since $(A_{-+}-A_{+-})(\omega)$ is an odd function of ω,

$$A(\zeta)=\int_0^{\infty}\frac{d\omega}{\pi}\frac{\omega(A_{-+}-A_{+-})(\omega)}{\omega^2-\zeta^2}.$$

We have already remarked on the generality of the representation

$$A_{-+}(\omega)/A_{+-}(\omega)=e^{\omega\beta(\omega)},\quad \beta(-\omega)=\beta(\omega),$$

and thus we shall write

$$(A_{-+}-A_{+-})(\omega)=a(\omega)\tanh[\tfrac{1}{2}\omega\beta(\omega)]$$
$$(A_{-+}+A_{+-})(\omega)=a(\omega)=a(-\omega)\geq 0,$$

which gives

$$G(\zeta)^{-1}=\omega_0^2-\zeta^2-\int_0^\infty \frac{d\omega}{\pi}\frac{\omega a(\omega)\tanh[\frac{1}{2}\omega\beta(\omega)]}{\omega^2-\zeta^2}.$$

Since this is an even function of ζ, it also represents the Fourier transform of G_a in the lower half-plane $\text{Im}\zeta<-\alpha$.

If the effective temperature is positive and finite at all frequencies, $\beta(\omega)>0$, $G(\zeta)$ can have no complex poles as a function of the variable ζ^2. A complex pole at $\zeta^2=x+iy$, $y\neq 0$, is a zero of $G(\zeta)^{-1}$ and requires that

$$y\left[1+\int_0^\infty \frac{d\omega}{\pi}\frac{\omega a(\omega)\tanh[\frac{1}{2}\omega\beta(\omega)]}{(\omega^2-x)^2+y^2}\right]=0,$$

which is impossible since the quantity in brackets exceeds unity. On letting y approach zero, we see that a pole of $G(\zeta)$ can occur at a point $x=\omega'^2>0$ only if $a(\omega')=0$. If the external system responds through the oscillator coupling to any impressed frequency, $a(\omega)>0$ for all ω and no pole can appear on the positive real axis of ζ^2. As to the negative real axis, $G(\zeta)^{-1}$ is a monotonically decreasing function of $\zeta^2=x$ that begins at $+\infty$ for $x=-\infty$ and will therefore have no zero on the negative real axis if it is still positive at $x=0$. The corresponding condition is

$$\omega_0^2>\int_0^\infty \frac{d\omega}{\pi}a(\omega)\frac{\tanh[\frac{1}{2}\omega\beta(\omega)]}{\omega}.$$

Under these circumstances $\alpha=0$, for $G(\zeta)$, *qua* function of ζ^2, has no singularity other than the branch line on the positive real axis, and the ζ singularities are therefore confined entirely to the real axis. This is indicated by

$$G(\zeta)=\int_0^\infty d\omega^2\frac{B(\omega^2)}{\omega^2-\zeta^2}=\int_{-\infty}^\infty d\omega\,\epsilon(\omega)\frac{B(\omega^2)}{\omega-\zeta},$$

and $B(\omega^2)$ is the positive quantity

$$B(\omega^2)=\frac{(2\pi)^{-1}a(\omega)\tanh[\frac{1}{2}|\omega|\beta(\omega)]}{\left[\omega_0^2-\omega^2-P\int_0^\infty \frac{d\omega'^2}{2\pi}\frac{\tanh(\frac{1}{2}\omega'\beta(\omega'))}{\omega'^2-\omega^2}a(\omega')\right]^2+[\frac{1}{2}a(\omega)\tanh\frac{1}{2}\omega\beta(\omega)]^2}.$$

Some integral relations are easily obtained by comparison of asymptotic forms. Thus

$$\int_0^\infty d\omega^2 B(\omega^2)=1,$$

$$\int_0^\infty d\omega^2\omega^2 B(\omega^2)=\omega_0^2,$$

and

$$\int_0^\infty d\omega^2\omega^4 B(\omega^2)=\omega_0^4+\int_0^\infty \frac{d\omega^2}{2\pi}a(\omega)\tanh[\tfrac{1}{2}\omega\beta(\omega)]$$
$$=\omega_0^4+\langle[i\dot{Q},Q]\rangle_\vartheta,$$

while setting $\zeta=0$ yields

$$\int_0^\infty d\omega^2\frac{B(\omega^2)}{\omega^2}=\left[\omega_0^2-\int_0^\infty \frac{d\omega}{\pi}a(\omega)\frac{\tanh\frac{1}{2}\omega\beta(\omega)}{\omega}\right]^{-1}.$$

The Green's functions are recovered on using the inverse Fourier transformation

$$G(t-t')=\int_{-\infty}^\infty \frac{d\zeta}{2\pi}e^{-i\zeta(t-t')}G(\zeta),$$

where the path of integration is drawn in the half-plane of regularity. Accordingly,

$$G_r(t-t')=\int_0^\infty d\omega^2 B(\omega^2)\frac{\sin\omega(t-t')}{\omega}\eta_+(t-t')$$

and

$$G_a(t-t')=-\int_0^\infty d\omega^2 B(\omega^2)\frac{\sin\omega(t-t')}{\omega}\eta_-(t-t').$$

The integral relations mentioned previously can be expressed in terms of these Green's functions. Thus,

$$\int_0^\infty d\tau G_r(\tau)=\left[\omega_0^2-\int_0^\infty \frac{d\omega}{\pi}a\frac{\tanh(\frac{1}{2}\omega\beta)}{\omega}\right]^{-1},$$

while, in the limit of small positive τ,

$$G_r(\tau)-(1/\omega_0)\sin\omega_0\tau\sim(\tau^5/5!)\langle[i\dot{Q},Q]\rangle_\vartheta,$$

which indicates the initial effect of the coupling to the external system.

The function $B(\omega^2)$ is bounded, and the Green's functions must therefore approach zero as $|t-t'|\to\infty$. Accordingly, all reference to the initial oscillator condition and to the time t_2 must eventually disappear. For sufficiently large $t-t_2$, $t'-t_2$, the function $w(t-t_2$,

$t'-t_2$) reduces to

$$w(t-t')=\int_{-\infty}^{\infty}d\tau d\tau' G_r(t-\tau)a(\tau-\tau')G_a(\tau'-t')$$

$$=\int_{-\infty}^{\infty}\frac{d\omega}{2\pi}e^{-i\omega(t-t')}G(\omega+i\epsilon)a(\omega)G(\omega-i\epsilon)|_{\epsilon\to 0}.$$

But

$$(1/2\pi)a(\omega)|G(\omega+i\epsilon)|^2=B(\omega^2)\coth[\tfrac{1}{2}|\omega|\beta(\omega)],$$

and, therefore,

$$w(t-t')=\int_0^{\infty}d\omega^2 B(\omega^2)\coth[\tfrac{1}{2}\omega\beta(\omega)]\frac{1}{\omega}\cos\omega(t-t').$$

The corresponding asymptotic form of the matrix $G(t-t_2, t'-t_2)$ is given by

$$G(t-t')=i\int_{-\infty}^{\infty}d\omega B(\omega^2)e^{-i\omega(t-t')}\begin{pmatrix}\eta_+(\omega)\eta_+(t-t')+\eta_-(\omega)\eta_-(t-t')+n, & -\eta_-(\omega)-n\\ -\eta_+(\omega)-n, & \eta_+(\omega)\eta_-(t-t')+\eta_-(\omega)\eta_+(t-t')+n\end{pmatrix},$$

with

$$n(\omega)=(e^{|\omega|\beta(\omega)}-1)^{-1},$$

which describes the oscillator in equilibrium at each frequency with the external system. When the temperature is frequency independent, this is thermal equilibrium. Note also that at zero temperature $n(\omega)=0$, and $G(t-t')_{++}$ is characterized by the temporal outgoing wave boundary condition—positive (negative) frequencies for positive (negative) time difference. The situation is similar for $G(t-t')_{--}$ as a function of $t'-t$.

It can no longer be maintained that placing $\beta_0=\beta$ removes all reference to the initial time. An interval must elapse before thermal equilibrium is established at the common temperature. This can be seen by evaluating the t_2 derivative of $w(t-t_2, t'-t_2)$:

$$\frac{\partial}{\partial t_2}w=-G_r(t-t_2)\int_{t_2}^{\infty}d\tau' a(t_2-\tau')G_a(\tau'-t')$$

$$-\int_{t_2}^{\infty}d\tau G_r(t-\tau)a(\tau-t_2)G_a(t_2-t')$$

$$+\frac{1}{\omega_0}\coth(\tfrac{1}{2}\omega_0\beta_0)$$

$$\times\left\{\frac{\partial}{\partial t_2}G_r(t-t_2)\int_{-\infty}^{\infty}d\tau' A_a(t_2-\tau')G_a(\tau'-t')\right.$$

$$\left.+\int_{-\infty}^{\infty}d\tau A_r(t-\tau)G_r(\tau-t_2)\frac{\partial}{\partial t_2}G_a(t_2-t')\right\},$$

for if this is to vanish, the integrals involving G_r, say, must be expressible as linear combinations of $G_r(t-t_2)$ and its time derivative, which returns us to the approximate treatment of the preceding section, including the approximate identification of ω_0 with the effective oscillator frequency. Hence $\vartheta_0=\vartheta$ does not represent the initial condition of thermal equilibrium between oscillator and external system. While it is perfectly clear that the latter situation is described by the matrix $G_\vartheta(t-t')$, a derivation that employs thermal equilibrium as an initial condition would be desirable.

The required derivation is produced by the device of computing the trace of the transformation function $\langle t_2'|t_2\rangle^{F_\pm}$, in which the return path terminates at the different time $t_2'=t_2-T$, and the external force $F_-(t)$ is zero in the interval between t_2 and t_2'. The particular significance of the trace appears on varying the parameter λ that measures the coupling between oscillator and external system:

$$\frac{\partial}{\partial\lambda}\langle t_2'|t_2\rangle^{F_\pm}=i\left\langle t_2'\left|\left[\int_{t_2}^{t_1}dtq_+Q_+(t)-\int_{t_2'}^{t_1}dtq_-Q_-(t)\right.\right.\right.$$

$$\left.\left.\left.+G_\lambda(t_2')-G_\lambda(t_2)\right]\right|t_2\right\rangle^{F_\pm}.$$

The operators G_λ are needed to generate infinitesimal transformations of the individual states at the corresponding times, if these states are defined by physical quantities that depend upon λ, such as the total energy. There is no analogous contribution to the trace, however, for the trace is independent of the representation, which is understood to be defined similarly at t_2 and t_2', and one could use a complete set that does not refer to λ. More generally, we observe that $G_\lambda(t_2')$ bears the same relation to the $\langle t_2'|$ states as does $G_\lambda(t_2)$ to the states at time t_2, and therefore

$$\mathrm{tr}\langle t_2'|G_\lambda(t_2')|t_2\rangle-\mathrm{tr}\langle t_2'|G_\lambda(t_2)|t_2\rangle=0.$$

Accordingly, the construction of an effective action operator can proceed as before, with appropriately modified ranges of time integration, and, for the external system, with

$$\langle Q(t)Q(t')\rangle=\frac{\mathrm{tr}\langle t_2'|Q(t)Q(t')|t_2\rangle}{\mathrm{tr}\langle t_2'|t_2\rangle}.$$

This trace structure implies that

$$\langle Q(t)Q(t_2)\rangle=\langle Q(t_2')Q(t)\rangle$$

or, since these correlation functions depend only on

time differences, that

$$A_{-+}(t-t_2)=A_{+-}(t-t_2'),$$

which is also expressed by

$$A_{-+}(\omega)=e^{-i\omega T}A_{+-}(\omega).$$

The equations of motion for $t>t_2$ are given by

$$\left(\frac{d^2}{dt^2}+\omega_0^2\right)(q_--q_+)(t)-\int_{-\infty}^{\infty}dt'A_a(t-t')(q_--q_+)(t')=(F_--F_+)(t)$$

and

$$\left(\frac{d^2}{dt^2}+\omega_0^2\right)(q_++q_-)(t)-\int_{t_2}^{\infty}dt'A_r(t-t')(q_++q_-)(t')+i\int_{t_2}^{\infty}dt'a(t-t')(q_--q_+)(t')=(F_++F_-)(t)-2i\int_{t_2'}^{t_2}dt'A_{+-}(t-t')q_-(t').$$

These are supplemented by the equation for $q_-(t)$ in the interval from t_2' to t_2:

$$\left(\frac{d^2}{dt^2}+\omega_0^2\right)q_-(t)+i\int_{t_2'}^{t_2}dt'A_{--}(t-t')q_-(t')=-i\int_{t_2}^{\infty}dt'A_{-+}(t-t')(q_--q_+)(t'),$$

and the effective boundary condition

$$q_-(t_2')=q_+(t_2).$$

The equation for q_--q_+ is solved as before,

$$(q_--q_+)(t)=\int_{-\infty}^{\infty}dt'G_a(t-t')(F_--F_+)(t'),$$

whereas

$$\begin{aligned}(q_++q_-)(t)&=\int_{-\infty}^{\infty}dt'G_r(t-t')(F_++F_-)(t')\\&-i\int_{t_2}^{\infty}d\tau G_r(t-\tau)\int_{t_2}^{\infty}dt'a(\tau-t')(q_--q_+)(t')\\&-2i\int_{t_2}^{\infty}d\tau G_r(t-\tau)\int_{t_2-T}^{t_2}dt'A_{+-}(\tau-t')q_-(t')\\&+G_r(t-t_2)\frac{\partial}{\partial t_2}(q_++q_-)(t_2)\\&-\frac{\partial}{\partial t_2}G_r(t-t_2)(q_++q_-)(t_2),\end{aligned}$$

which has been written for external forces that are zero until the moment t_2 has passed.

Perhaps the simplest procedure at this point is to ask for the dependence of the latter solution upon t_2, for fixed T. We find that

$$\begin{aligned}\frac{\partial}{\partial t_2}(q_++q_-)(t)=&-\int_{t_2}^{\infty}dt'G_r(t-t')A_r(t'-t_2)(q_++q_-)(t_2)\\&+i\int_{t_2}^{\infty}dt'G_r(t-t')a(t'-t_2)(q_--q_+)(t_2)\\&-2i\int_{t_2}^{\infty}dt'G_r(t-t')[A_{+-}(t'-t_2)q_-(t_2)\\&\qquad-A_{-+}(t'-t_2)q_+(t_2)],\end{aligned}$$

on using the relations

$$\int_{-\infty}^{\infty}d\tau A_r(t-\tau)G_r(\tau-t')=\int_{-\infty}^{\infty}d\tau G_r(t-\tau)A_r(\tau-t'),$$

$$A_{+-}(t-t_2')q_-(t_2')=A_{-+}(t-t_2)q_+(t_2).$$

Therefore,

$$(\partial/\partial t_2)(q_++q_-)(t)=0,$$

since, with positive time argument,

$$a-iA_r=2A_{-+}$$
$$a+iA_r=2A_{+-}.$$

The utility of this result depends upon the approach of the Green's functions to zero with increasing magnitude of the time argument, which is assured, after making the substitution $T\rightarrow i\beta$, under the circumstances we have indicated. Then we can let $t_2\rightarrow-\infty$ and obtain

$$(q_++q_-)(t)=\int_{-\infty}^{\infty}dt'G_r(t-t')(F_++F_-)(t')-i\int_{-\infty}^{\infty}dt'w(t-t')(F_--F_+)(t')$$

with

$$w(t-t')=\int_{-\infty}^{\infty}d\tau d\tau' G_r(t-\tau)a(\tau-\tau')G_a(\tau'-t'),$$

as anticipated.

Our results determine the trace ratio

$$\frac{\mathrm{tr}\langle t_2'|t_2\rangle^{F\pm}}{\mathrm{tr}\langle t_2'|t_2\rangle}=\frac{\mathrm{tr}\langle t_2|e^{iTH}|t_2\rangle^{F\pm}}{\mathrm{tr}e^{iTH}},$$

where H is the Hamiltonian operator of the complete system, and the substitution $T\rightarrow i\beta$ yields the transformation function

$$\langle t_2|t_2\rangle_\beta{}^{F\pm}=\exp\left[\tfrac{1}{2}i\int dtdt'F(t)G_\beta(t-t')F(t')\right]$$

with

$$G_\vartheta(t-t')=\tfrac{1}{2}G_r(t-t')\begin{pmatrix}1 & 1\\ -1 & -1\end{pmatrix}+\tfrac{1}{2}G_a(t-t')\begin{pmatrix}1 & -1\\ 1 & -1\end{pmatrix}+\tfrac{1}{2}iw(t-t')\begin{pmatrix}1 & -1\\ -1 & 1\end{pmatrix}$$

and

$$w(t-t')=\int_{-\infty}^{\infty}d\omega B(\omega^2)\coth(\tfrac{1}{2}|\omega|\beta)e^{-i\omega(t-t')}.$$

We can also write

$$w(t-t')=\int_{-\infty}^{\infty}d\tau C(t-\tau)(G_a-G_r)(\tau-t'),$$

where

$$C(t-t')=\frac{i}{2\pi}P\int_{-\infty}^{\infty}d\omega\coth(\tfrac{1}{2}\omega\beta)e^{-i\omega(t-t')}=\frac{1}{\beta}\coth\left[\frac{\pi}{\beta}(t-t')\right].$$

What is asserted here about expectation values in the presence of an external field $F(t)$ becomes explicit on writing

$$F_\pm(t)=f_\pm(t)+F(t)$$

and indicating the effect of $f_\pm(t)$ by equivalent time-ordered operators,

$$\left\langle\left(\exp\left[-i\int dtf_-(t)q(t)\right]\right)_-\times\left(\exp\left[i\int dtf_+(t)q(t)\right]\right)_+\right\rangle_\vartheta^F$$

$$=\exp\left\{\tfrac{1}{2}i\int dtdt'f(t)G_\vartheta(t-t')f(t')+i\int dtdt'(f_+-f_-)(t)G_r(t-t')F(t')\right\}.$$

Thus

$$\langle q(t)\rangle_\vartheta{}^F=\int_{-\infty}^{\infty}dt'G_r(t-t')F(t')$$

and the properties of $q-\langle q\rangle_\vartheta{}^F$, which are independent of F, are given by setting $F=0$ in the general result. In particular, we recover the matrix identity

$$G_\vartheta(t-t')=i\begin{pmatrix}\langle(q(t)q(t'))_+\rangle_\vartheta, & -\langle q(t')q(t)\rangle_\vartheta\\ \langle q(t)q(t')\rangle_\vartheta, & \langle(q(t)q(t'))_-\rangle_\vartheta\end{pmatrix}.$$

The relation between w and G_a-G_r can then be displayed as a connection between symmetrical product and commutator expectation values

$$\langle\{q(t),q(t')\}\rangle_\vartheta=\int_{-\infty}^{\infty}d\tau C(t-\tau)\left\langle\frac{1}{i}[q(\tau),q(t')]\right\rangle_\vartheta.$$

In addition to the trace ratio, which determines the thermal average transformation function $\langle t_2|t_2\rangle_\vartheta{}^{F\pm}$ with its attendant physical information, it is possible to compute the trace

$$\mathrm{tr}\langle t_2'|t_2\rangle=\mathrm{tr}e^{iTH}\to\mathrm{tr}e^{-\beta H}$$

which describes the complete energy spectrum and thereby the thermostatic properties of the oscillator in equilibrium with the external system. For this purpose we set $F_\pm=0$ for $t>t_2$ and apply an arbitrary external force $F_-(t)$ in the interval from t_2' to t_2. Moreover, the coupling term between oscillator and external system in the effective action operator is supplied with the variable factor λ (formerly λ^2). Then we have

$$\frac{\partial}{\partial\lambda}\mathrm{tr}\langle t_2'|t_2\rangle^{F-}=-\tfrac{1}{2}\,\mathrm{tr}\left\langle t_2'\left|\int_{t_2'}^{t_2}dtdt'A_{--}(t-t')(q(t)q(t'))_-\right|t_2\right\rangle^{F-}$$

$$=-\tfrac{1}{2}i\int_{t_2'}^{t_2}dtdt'A_{--}(t-t')\frac{\delta}{\delta F_-(t')}\mathrm{tr}\langle t_2'|q_-(t)|t_2\rangle^{F-},$$

where $q_-(t)$ obeys the equation of motion

$$\left(\frac{d^2}{dt^2}+\omega_0{}^2\right)q_-(t)+i\lambda\int_{t_2'}^{t_2}dt'A_{--}(t-t')q_-(t')=F_-(t)$$

with the accompanying boundary condition

$$q_-(t_2')=q_+(t_2)=q_-(t_2),$$

which is a statement of periodicity for the interval $T=t_2'-t_2$. The solution of this equation is

$$q_-(t)=\int_{t_2'}^{t_2}dt'G(t-t')F_-(t'),$$

where the Green's function obeys

$$\left(\frac{d^2}{dt^2}+\omega_0{}^2\right)G(t-t')+i\lambda\int_{t_2'}^{t_2}d\tau A_{--}(t-\tau)G(\tau-t')=\delta(t-t')$$

and the requirement of periodicity. We can now place $F_-=0$ in the differential equation for the trace, and obtain

$$\frac{\partial}{\partial\lambda}\log\mathrm{tr}\langle t_2'|t_2\rangle=-\tfrac{1}{2}i\int_{t_2'}^{t_2}dtdt'A_{--}(t-t')G(t-t').$$

The periodic Green's function is given by the Fourier series

$$G(t-t')=\frac{1}{T}\sum_{n=-\infty}^{\infty}\exp\left[-\frac{2\pi in}{T}(t-t')\right]G(n)$$

with

$$G(n)=\left[\omega_0{}^2-\left(\frac{2\pi n}{T}\right)^2-\lambda A(n)\right]^{-1}=G(-n)$$

and

$$A(n)=-\int_{t_2'}^{t_2} dt \exp\left[\frac{2\pi in}{T}(t-t')\right] iA_{--}(t-t')$$

$$=\int_0^\infty \frac{\omega d\omega}{\pi} \frac{(A_{-+}-A_{+-})(\omega)}{\omega^2-(2\pi n/T)^2},$$

where, it is to be recalled,

$$A_{-+}(\omega)=e^{-i\omega T}A_{+-}(\omega),$$

so that the integrand has no singularities at $\omega T=2\pi|n|$. Now we have

$$\frac{\partial}{\partial\lambda}\log \text{tr}=\tfrac{1}{2}\sum_{-\infty}^{\infty} A(n)G(n)$$

$$=-\frac{1}{2}\frac{\partial}{\partial\lambda}\sum_{-\infty}^{\infty}\log\left[\omega_0^2-\left(\frac{2\pi n}{T}\right)^2-\lambda A(n)\right]$$

which, together with the initial condition

$$\lambda=0:\quad \text{tr}e^{iTH}=(\text{tr}_e e^{iTH_{\text{ext}}})\sum_{n=0}^{\infty} e^{i(n+\frac{1}{2})\omega_0 T}$$

$$=(\text{tr}_e)(i/2\sin\tfrac{1}{2}\omega_0 T),$$

yields

$$\text{tr}e^{iTH}=(\text{tr}_e)(i/2\sin\tfrac{1}{2}\omega_0 T)$$

$$\times\exp\left\{-\tfrac{1}{2}\sum_{-\infty}^{\infty}\log\left[\frac{\omega_0^2-(2\pi n/T)^2-A(n)}{\omega_0^2-(2\pi n/T)^2}\right]\right\}.$$

We have already introduced the function

$$G^{-1}(\zeta)=\omega_0^2-\zeta^2-\int_0^\infty\frac{\omega d\omega}{\pi}\frac{(A_{-+}-A_{+-})(\omega)}{\omega^2-\zeta^2}$$

and examined some of its properties for real and positive $A_{-+}(\omega)$, $A_{+-}(\omega)$. This situation is recovered on making the substitution $T\to i\beta$, and thus

$$Z=\text{tr}e^{-\beta H}=Z_e(1/2\sinh\tfrac{1}{2}\beta\omega_0)$$

$$\times\exp\left\{-\tfrac{1}{2}\sum_{-\infty}^{\infty}\log\left[\frac{G^{-1}(i2\pi n/\beta)}{\omega_0^2+(2\pi n/\beta)^2}\right]\right\},$$

the existence of which for all $\beta>0$ requires that $G^{-1}(\zeta)$ remain positive at every value comprised in $\zeta^2=-(2\pi n/\beta)^2$, which is to say the entire negative ζ^2 axis including the origin. The condition

$$G^{-1}(0)>0$$

is thereby identified as a stability criterion. To evaluate the summation over n most conveniently we shall give an alternative construction for the function $\log(G^{-1}(\zeta)/-\zeta^2)$, which, as a function of ζ^2, has all its singularities located on the branch line extending from 0 to ∞ and vanishes at infinity in this cut plane. Hence

$$\log(G^{-1}(\zeta)/-\zeta^2)=\frac{1}{\pi}\int_0^\infty d\omega^2\frac{\varphi(\omega)}{\omega^2-\zeta^2},$$

where the value

$$\varphi(0)=\pi$$

reproduces the pole of $G^{-1}(\zeta)/(-\zeta^2)$ at $\zeta^2=0$. We also recognize, on relating the two forms,

$$G^{-1}(\zeta)=(-\zeta^2)\exp\left[\frac{1}{\pi}\int_0^\infty d\omega^2\frac{\varphi(\omega)}{\omega^2-\zeta^2}\right]$$

$$=\omega_0^2-\zeta^2-\int_0^\infty\frac{d\omega^2}{2\pi}\frac{a(\omega)\tanh(\frac{1}{2}\omega\beta)}{\omega^2-\zeta^2},$$

that

$$-\tfrac{1}{2}a(\omega)\tanh(\tfrac{1}{2}\omega\beta)\cot\varphi(\omega)$$

$$=\omega_0^2-\omega^2-P\int_0^\infty\frac{d\omega'}{\pi}\frac{\omega' a(\omega')\tanh(\frac{1}{2}\omega'\beta)}{\omega'^2-\omega^2}.$$

The positive value of the right-hand side as $\omega\to 0$ shows that $\varphi(\omega)$ approaches the zero frequency limiting value of π from below, and the assumption that $a(\omega)>0$ for all ω implies

$$\pi\geq\varphi(\omega)>0,$$

where the lower limit is approached as $\omega\to\infty$.

A comparison of asymptotic forms for $G^{-1}(\zeta)$ shows that

$$\omega_0^2=\frac{1}{\pi}\int_0^\infty d\omega^2\varphi(\omega)=\int_0^\infty d\omega\left(-\frac{1}{\pi}\frac{d\varphi(\omega)}{d\omega}\right)\omega^2,$$

while

$$\int_0^\infty d\omega\left(-\frac{1}{\pi}\frac{d\varphi(\omega)}{d\omega}\right)\omega^4=\omega_0^4+2\langle[i\dot{Q},Q]\rangle_\beta.$$

The introduction of the phase derivative can also be performed directly in the structure of $G^{-1}(\zeta)$,

$$G^{-1}(\zeta)=\exp\left[\int_0^\infty d\omega\left(-\frac{1}{\pi}\frac{d\varphi(\omega)}{d\omega}\right)\log(\omega^2-\zeta^2)\right],$$

and equating the two values for $G^{-1}(0)$ gives

$$\int_0^\infty d\omega\left(-\frac{1}{\pi}\frac{d\varphi(\omega)}{d\omega}\right)\log\omega^2$$

$$=\log\left[\omega_0^2-\int_0^\infty\frac{d\omega}{\pi}a(\omega)\frac{\tanh\frac{1}{2}\omega\beta}{\omega}\right].$$

We now have the representation

$$\log\left[\frac{G^{-1}(i2\pi n/\beta)}{\omega_0^2+(2\pi n/\beta)^2}\right]$$

$$=\int_0^\infty d\omega\left(-\frac{1}{\pi}\frac{d\varphi(\omega)}{d\omega}\right)\log\frac{\omega^2+(2\pi n/\beta)^2}{\omega_0^2+(2\pi n/\beta)^2},$$

and the summation formula derived from the product form of the hyperbolic sine function,

$$\frac{1}{2}\sum_{-\infty}^{\infty}\log\frac{\omega^2+(2\pi n/\beta)^2}{\omega_0^2+(2\pi n/\beta)^2}=\log\left[\frac{\sinh\frac{1}{2}\omega\beta}{\sinh\frac{1}{2}\omega_0\beta}\right],$$

gives us the desired result

$$Z=Z_e\exp\left[-\int_0^\infty d\omega\left(-\frac{1}{\pi}\frac{d\varphi(\omega)}{d\omega}\right)\log 2\sinh\tfrac{1}{2}\omega\beta\right].$$

The second factor can be ascribed to the oscillator, with its properties modified by interaction with the external system. The average energy of the oscillator at temperature $\vartheta=\beta^{-1}$ is therefore given by

$$E=\frac{\partial}{\partial\beta}\int_0^\infty d\omega\left(-\frac{1}{\pi}\frac{d\varphi}{d\omega}\right)\log 2\sinh\tfrac{1}{2}\omega\beta$$

in which the temperature dependence of the phase $\varphi(\omega)$ is not to be overlooked. In an extreme high-temperature limit, such that $\omega\beta\ll 1$ for all significant frequencies, we have

$$E\simeq\frac{\partial}{\partial\beta}[\log\beta+\tfrac{1}{2}\log(\omega_0^2-\beta\langle Q^2\rangle_\vartheta)],$$

and the simple classical result $E=\vartheta$ appears when $\langle Q^2\rangle_\vartheta$ is proportional to ϑ. The oscillator energy at zero temperature is given by

$$E_0=\tfrac{1}{2}\int_0^\infty d\omega\left(-\frac{1}{\pi}\frac{d\varphi}{d\omega}\right)_{\vartheta=0}\omega,$$

and the oscillator contribution to the specific heat vanishes.

The following physical situation has consequences that resemble the simple model of the previous section. For values of $\omega\lesssim\omega_0$, $a(\omega)\tanh(\frac{1}{2}\omega\beta)\ll\omega_0^2$, and $a(\omega)$ differs significantly from zero until one attains frequencies that are large in comparison with ω_0. The magnitudes that $a(\omega)$ can assume at frequencies greater than ω_0 is limited only by the assumed absence of rapid variations and by the requirement of stability. The latter is generally assured if

$$\frac{1}{\pi}\int_{\sim\omega_0}^\infty\frac{d\omega}{\omega}a(\omega)<\omega_0^2.$$

We shall suppose that the stability requirement is comfortably satisfied, so that the right-hand side of the equation for $\cot\varphi(\omega)$ is an appreciable fraction of ω_0^2 at sufficiently low frequencies. Then $\tan\varphi$ is very small at such frequencies, or $\varphi(\omega)\sim\pi$, and this persists until we reach the immediate neighborhood of the frequency $\omega_1<\omega_0$ such that

$$\omega_0^2-\omega_1^2-P\int_0^\infty\frac{d\omega}{\pi}\frac{\omega a(\omega)\tanh(\frac{1}{2}\omega\beta)}{\omega^2-\omega_1^2}=0.$$

That the function in question, $\mathrm{Re}G^{-1}(\omega+i0)$, has a zero, follows from its positive value at $\omega=0$ and its asymptotic approach to $-\infty$ with indefinitely increasing frequency. Under the conditions we have described, with the major contribution to the integral coming from high frequencies, the zero point is given approximately as

$$\omega_1^2\simeq\omega_0^2-\int_0^\infty\frac{d\omega}{\pi}a(\omega)\frac{\tanh(\frac{1}{2}\omega\beta)}{\omega},$$

and somewhat more accurately by

$$\omega_1^2=B\left[\omega_0^2-\int_0^\infty\frac{d\omega}{\pi}a(\omega)\frac{\tanh(\frac{1}{2}\omega\beta)}{\omega}\right],$$

where

$$B^{-1}=1+P\int_0^\infty\frac{d\omega}{2\pi}\frac{1}{\omega^2-\omega_1^2}\frac{d}{d\omega}[a(\omega)\tanh(\tfrac{1}{2}\omega\beta)].$$

As we shall see, B is less than unity, but only slightly so under the circumstances assumed.

In the neighborhood of the frequency ω_1, the equation that determines $\varphi(\omega)$ can be approximated by

$$-\tfrac{1}{2}a(\omega_1)\tanh(\tfrac{1}{2}\omega_1\beta)\cot\varphi(\omega)=B^{-1}(\omega_1^2-\omega^2)$$

or

$$\cot\varphi(\omega)=(\omega^2-\omega_1^2)/\gamma\omega_1\simeq(\omega-\omega_1)/\tfrac{1}{2}\gamma,$$

with the definition

$$\gamma=\tfrac{1}{2}Ba(\omega_1)[\tanh(\tfrac{1}{2}\omega_1\beta)/\omega_1]\ll\omega_1.$$

Hence, as ω rises through the frequency ω_1, φ decreases abruptly from a value close to π to one near zero. The subsequent variations of the phase are comparatively gradual, and φ eventually approaches zero as $\omega\to\infty$. A simple evaluation of the average oscillator energy can be given when the frequency range $\omega>\omega_1$ over which $a(\omega)$ is appreciable in magnitude is such that $\beta\omega\gg 1$. There will be no significant temperature variation in the latter domain and in particular ω_1 should be essentially temperature independent. Then, since $-(1/\pi)(d\varphi/d\omega)$ in the neighborhood of ω_1 closely resembles $\delta(\omega-\omega_1)$, we have approximately

$$E=\frac{\partial}{\partial\beta}\left[\log(2\sinh\tfrac{1}{2}\omega_1\beta)+\beta\int_{>\omega_1}^\infty d\omega\left(-\frac{1}{\pi}\frac{d\varphi}{d\omega}\right)\tfrac{1}{2}\omega\right]$$

$$=\omega_1\left(\frac{1}{e^{\beta\omega_1}-1}+\tfrac{1}{2}\right)+\int_{>\omega_1}^\infty d\omega\left(-\frac{1}{\pi}\frac{d\varphi}{d\omega}\right)\tfrac{1}{2}\omega,$$

which describes a simple oscillator of frequency ω_1, with a displaced origin of energy.

Note that with $\varphi(\omega)$ very small at a frequency slightly greater than ω_1 and zero at infinite frequency, we have

$$\int_{>\omega_1}^\infty d\omega\left(-\frac{1}{\pi}\frac{d\varphi}{d\omega}\right)\tfrac{1}{2}\omega\simeq\frac{1}{2\pi}\int_{>\omega_1}^\infty d\omega\,\varphi(\omega)>0.$$

Related integrals are

$$\omega_0^2-\omega_1^2 \simeq \frac{2}{\pi}\int_{>\omega_1}^{\infty} d\omega\omega\varphi(\omega)>0$$

and

$$\log B^{-1} \simeq \frac{2}{\pi}\int_{>\omega_1}^{\infty} \frac{d\omega}{\omega}\varphi(\omega)>0.$$

The latter result confirms that $B<1$. A somewhat more accurate formula for B is

$$B=\exp\left[-\int_{>\omega_1}^{\infty} d\omega\left(-\frac{1}{\pi}\frac{d\varphi(\omega)}{d\omega}\right)\log(\omega^2-\omega_1^2)\right].$$

If the major contributions to all these integrals come from the general vicinity of a frequency $\bar{\omega}\gg\omega_0$, we can make the crude estimates

$$\frac{1}{2\pi}\int_{>\omega_1}^{\infty} d\omega\varphi(\omega)\sim\frac{\omega_0^2}{\bar{\omega}}\ll\omega_1,\ \log B^{-1}\sim\left(\frac{\omega_0}{\bar{\omega}}\right)^2\ll 1.$$

Then neither the energy shift nor the deviation of the factor B from unity are particularly significant effects.

The approximation of $\text{Re}G(\omega+i0)$ as $B^{-1}(\omega_1^2-\omega^2)$ evidently holds from zero frequency up to a frequency considerably in excess of ω_1. Throughout this frequency range we have

$$-\tfrac{1}{2}a(\omega)\tanh(\tfrac{1}{2}\omega\beta)\cot\varphi(\omega)=B^{-1}(\omega_1^2-\omega^2)$$

or

$$\cot\varphi(\omega)=(\omega_1^2-\omega^2)/\gamma\omega$$

with

$$\gamma(\omega)=\tfrac{1}{2}Ba(\omega)\tanh(\tfrac{1}{2}\omega\beta)/\omega.$$

If in particular $\beta\omega_1\ll 1$, the frequencies under consideration are in the classical domain and γ is the frequency independent constant

$$\gamma=\tfrac{1}{4}Ba(0)\beta.$$

To regard γ as constant for a quantum oscillator requires a suitable frequency restriction to the vicinity of ω_1. The function $B(\omega^2)$ can be computed from

$$B(\omega^2)=\frac{2}{\pi}\frac{\sin^2\varphi(\omega)}{a(\omega)\tanh(\tfrac{1}{2}\omega\beta)}$$

$$=\frac{1}{\pi}\frac{B}{\gamma\omega}\frac{1}{\cot^2\varphi+1},$$

and accordingly is given by

$$B(\omega^2)=B\frac{1}{\pi}\frac{\gamma\omega}{(\omega^2-\omega_1^2)^2+(\gamma\omega)^2}$$

$$=B\frac{\gamma\omega}{\pi}\left|\frac{1}{\omega^2+i\gamma\omega-\omega_1^2}\right|^2.$$

The further concentration on the immediate vicinity of ω_1, $|\omega-\omega_1|\sim\gamma$, gives

$$B(\omega^2)=\frac{B}{2\omega_1}\frac{1}{\pi}\frac{\tfrac{1}{2}\gamma}{(\omega-\omega_1)^2+(\tfrac{1}{2}\gamma)^2}$$

which clearly identifies $B<1$ with the contribution to the integral $\int d\omega^2 B(\omega^2)$ that comes from the vicinity of this resonance of width γ at frequency ω_1, although the same result is obtained without the last approximation. The remainder of the integral, $1-B$, arises from frequencies considerably higher than ω_1 according to our assumptions.

There is a similar decomposition of the expressions for the Green's functions. Thus, with $t>t'$,

$$G_r(t-t')\simeq B\int_{-\infty}^{\infty}\frac{d\omega}{2\pi}e^{-i\omega(t-t')}\frac{1}{-\omega^2-i\gamma\omega+\omega_1^2}$$

$$+\int_{\gg\omega_1^2}^{\infty} d\omega^2 B(\omega^2)\frac{\sin\omega(t-t')}{\omega}.$$

The second high-frequency term will decrease very quickly on the time scale set by $1/\omega_1$. Accordingly, in using this Green's function, say in the evaluation of

$$\langle q(t)\rangle_\beta{}^F=\int_{-\infty}^{\infty} dt' G_r(t-t')F(t')$$

for an external force that does not vary rapidly in relation to ω_1, the contribution of the high-frequency term is essentially given by

$$F(t)\int_0^{\infty} d(t-t')\int_{\gg\omega_1^2}^{\infty} d\omega^2 B(\omega^2)\frac{\sin\omega(t-t')}{\omega}$$

$$=F(t)\int_{\gg\omega_1^2}^{\infty} d\omega^2\frac{B(\omega^2)}{\omega^2}.$$

But

$$\int_{\gg\omega_1^2}^{\infty} d\omega^2\frac{B(\omega^2)}{\omega^2}\simeq\left[\omega_0^2-\int_0^{\infty}\frac{d\omega}{\pi}a(\omega)\frac{\tanh\tfrac{1}{2}\omega\beta}{\omega}\right]^{-1}-\frac{B}{\omega_1^2},$$

$$\simeq 0,$$

and the response to such an external force is adequately described by the low-frequency part of the Green's function. We can represent this situation by an equivalent differential equation

$$\left(\frac{d^2}{dt^2}+\gamma\frac{d}{dt}+\omega_1^2\right)\langle q(t)\rangle_\beta{}^F=BF(t)$$

which needs no further qualification when the oscillations are classical but implies a restriction to a frequency interval within which γ is constant, for quantum oscillations. We note the reduction in the effectiveness of the external force by the factor B. Under the circumstances outlined this effect is not important and we shall place B equal to unity.

One can make a general replacement of the Green's functions by their low-frequency parts:

$$G_r(t-t') \to e^{-\frac{1}{2}\gamma(t-t')}\frac{1}{\omega_1}\sin(\omega_1(t-t'))\eta_+(t-t')$$

$$G_a(t-t') \to -e^{-\frac{1}{2}\gamma|t-t'|}\frac{1}{\omega_1}\sin(\omega_1(t-t'))\eta_-(t-t'),$$

if one limits the time localizability of measurements so that only time averages of $q(t)$ are of physical interest. This is represented in the expectation value formula by considering only functions $f_\pm(t)$ that do not vary too quickly. The corresponding replacement for $w(t-t')$ is

$$w(t-t') \to \coth(\tfrac{1}{2}\omega_1\beta)e^{-\frac{1}{2}\gamma|t-t'|}\frac{1}{\omega_1}\cos\omega_1(t-t'),$$

and the entire matrix $G_\vartheta(t-t')$ obtained in this way obeys the differential equation

$$\left(\frac{d^2}{dt^2}+\gamma\frac{d}{dt}+\omega_1^2\right)\left(\frac{d^2}{dt'^2}+\gamma\frac{d}{dt'}+\omega_1^2\right)G_\vartheta(t-t')$$

$$=\begin{pmatrix}1 & 0\\ 0 & -1\end{pmatrix}\left(\frac{d^2}{dt^2}+\omega_1^2\right)\delta(t-t')$$

$$+\begin{pmatrix}0 & -1\\ 1 & 0\end{pmatrix}\gamma\frac{d}{dt}\delta(t-t')+\begin{pmatrix}1 & -1\\ -1 & 1\end{pmatrix}\tfrac{1}{2}ia\delta(t-t'),$$

where $a=a(\omega_1)$.

The simplest presentation of results is again to be found in the Langevin viewpoint, which directs the emphasis from the coordinate operator $q(t)$ to the fluctuating force defined by

$$\left(\frac{d^2}{dt^2}+\gamma\frac{d}{dt}+\omega_1^2\right)q(t)=F(t)+F_f(t),$$

which is to say

$$F_f(t)=Q(t)+\gamma\frac{d}{dt}q(t)+(\omega_1^2-\omega_0^2)q(t).$$

This change is introduced by the substitution

$$f_\pm(t)=\left(\frac{d^2}{dt^2}-\gamma\frac{d}{dt}+\omega_1^2\right)k_\pm(t),$$

and the necessary partial integrations involve the previously established lemma on time-ordered operators, which here asserts that

$$\left(\exp\left[i\int dtfq\right]\right)_+=\left(\exp\left[i\int dtk(F+F_f)\right]\right)_+$$

$$\times\exp\left\{\tfrac{1}{2}i\int dt[(\omega_1^2-\omega_0^2)k^2+\omega_1^2k^2-(dk/dt)^2]\right\}.$$

We now find

$$\left\langle\left(\exp\left[-i\int dtk_-F_f\right]\right)_-\left(\exp\left[i\int dtk_+F_f\right]\right)_+\right\rangle_\vartheta$$

$$=\exp\left[-\tfrac{1}{2}\int dtdt'k(t)\zeta(t-t')k(t')\right]$$

with

$$\zeta(t-t')=\tfrac{1}{2}a\begin{pmatrix}1 & -1\\ -1 & 1\end{pmatrix}\delta(t-t')$$

$$-i(\omega_0^2-\omega_1^2)\begin{pmatrix}1 & 0\\ 0 & -1\end{pmatrix}\delta(t-t')$$

$$+i\gamma\begin{pmatrix}0 & 1\\ -1 & 0\end{pmatrix}\frac{d}{dt}\delta(t-t').$$

The latter matrix can also be identified as

$$\zeta(t-t')=\begin{pmatrix}\langle(F_f(t)F_f(t'))_+\rangle_\vartheta, & -\langle F_f(t')F_f(t)\rangle_\vartheta\\ -\langle F_f(t)F_f(t')\rangle_\vartheta, & \langle(F_f(t)F_f(t'))_-\rangle_\vartheta\end{pmatrix}.$$

In the classical limit

$$\left\langle\exp\left[i\int dtkF_f\right]\right\rangle_\vartheta=\exp\left[-\tfrac{1}{4}a\int dtk^2\right]$$

and

$$\tfrac{1}{4}a=\gamma\vartheta.$$

If a comparison is made with the similar results of the previous section it can be appreciated that the frequency range has been extended and the restriction $\omega_1\simeq\omega_0$ removed.

We return from these extended considerations on thermal equilibrium and consider one extreme example of negative temperature for the external system. This is described by

$$a(\omega)=a\delta(\omega-\omega_1),\quad \omega>0$$

and

$$-\beta(\omega_1)=|\beta|>0.$$

With the definition

$$(1/\pi)\omega_1a\tanh(\tfrac{1}{2}\omega_1|\beta|)=(\omega_1\mu)^2,$$

we have

$$G(\zeta)=\left[\omega_0^2-\zeta^2+\frac{(\omega_1\mu)^2}{\omega_1^2-\zeta^2}\right]^{-1}$$

$$=\frac{\omega_1^2-\zeta^2}{[\zeta^2-\frac{1}{2}(\omega_0^2+\omega_1^2)]^2+(\omega_1\mu)^2-[\frac{1}{2}(\omega_0^2-\omega_1^2)]^2}.$$

As a function of ζ^2, $G(\zeta)$ now has complex poles if

$$\tfrac{1}{2}|\omega_0^2-\omega_1^2|<\omega_1\mu.$$

We shall suppose, for simplicity, that $\omega_1=\omega_0$ and $\mu\ll\omega_0$. Then the poles of

$$G(\zeta)=\tfrac{1}{2}[(\omega_0^2+i\omega_0\mu-\zeta^2)^{-1}+(\omega_0^2-i\omega_0\mu-\zeta^2)^{-1}]$$

are located at $\zeta=\pm(\omega_0+\frac{1}{2}i\mu)$ and $\zeta=\pm(\omega-\frac{1}{2}i\mu)$. Accordingly, $G(\zeta)$ is regular outside a strip of width $2\alpha=\mu$. The associated Green's functions are given by

$$G_r(t-t')=\cosh(\tfrac{1}{2}\mu(t-t'))(\omega_0)^{-1}\sin(\omega_0(t-t'))\eta_+(t-t'),$$
$$G_a(t-t')=-\cosh(\tfrac{1}{2}\mu(t-t'))(\omega_0)^{-1}\sin(\omega_0(t-t'))\eta_-(t-t'),$$

and the function $w(t-t_2,\,t'-t_2)$, computed for $\omega_0(t-t_2)$, $\omega_0(t'-t_2)\gg1$, is

$$\begin{aligned}w(t-t_2,\,t'-t_2)&\\ \simeq(\omega_0)^{-1}\cos(\omega_0(t-t'))&[\coth(\tfrac{1}{2}\omega_0|\beta|)\sinh(\tfrac{1}{2}\mu(t-t_2))\\ &\times\sinh(\tfrac{1}{2}\mu(t'-t_2))+\coth(\tfrac{1}{2}\omega_0\beta_0)\\ &\times\cosh(\tfrac{1}{2}\mu(t-t_2))\cosh(\tfrac{1}{2}\mu(t'-t_2))].\end{aligned}$$

After the larger time intervals $\mu(t-t_2)$, $\mu(t'-t_2)\gg1$, we have

$$\begin{aligned}w(t-t_2,\,t'-t_2)\sim(2\omega_0)^{-1}e^{+\frac{1}{2}\mu(t-t_2)}e^{+\frac{1}{2}\mu(t'-t_2)}\cos\omega_0(t-t')&\\ \times[n_0+(1-e^{-\omega_0|\beta|})^{-1}],&\end{aligned}$$

with

$$n_0=(e^{\omega_0\beta_0}-1)^{-1}.$$

When t is in the vicinity of a time t_1, such that the amplification factor

$$k\simeq\tfrac{1}{2}e^{\frac{1}{2}\mu(t_1-t_2)}\gg1,$$

the oscillator is described by the classical coordinate

$$q(t)=k[q_s(t)+q_n(t)].$$

Here

$$q_s(t)=\int_{t_2}^{\infty}dt'\frac{1}{\omega_0}\sin(\omega_0(t-t'))F(t')e^{-\frac{1}{2}\mu(t'-t_2)}$$

and

$$q_n(t)=q_1\cos\omega t+q_2\sin\omega t,$$

where q_1 and q_2 are characterized by the expectation value formula

$$\langle e^{i(q_1f_1+q_2f_2)}\rangle=\exp[-(\nu/\omega_0)\tfrac{1}{2}(f_1^2+f_2^2)],$$

in which

$$\nu=n_0+(1-e^{-\omega_0|\beta|})^{-1}.$$

Accordingly, the probability of observing q_1 and q_2 within the range dq_1, dq_2 is

$$p(q_1q_2)dq_1dq_2=\frac{1}{2\pi}\frac{\omega_0}{\nu}\exp\left[-\frac{\omega_0}{\nu}\frac{1}{2}(q_1^2+q_2^2)\right]dq_1dq_2$$

$$=\frac{\omega_0}{\nu}\exp\left(-\frac{1}{2}\frac{\omega_0}{\nu}q_n^2\right)q_ndq_n\frac{1}{2\pi}d\varphi,$$

where q_n and φ are the amplitude and phase of $q_n(t)$. Despite rather different assumptions about the external system, these are the same conclusions as before, apart from a factor of $\frac{1}{2}$ in the formula for the gain.

GENERAL THEORY

The whole of the preceding discussion assumes an external system that is only slightly influenced by the presence of the oscillator. Now we must attempt to place this simplification within the framework of a general formulation. A more thorough treatment is also a practical necessity in situations such as those producing amplification of the oscillator motion, for a sizeable reaction in the external system must eventually appear, unless a counter mechanism is provided.

It is useful to supplement the previous Lagrangian operator with the term $q'(t)Q$, in which $q'(t)$ is an arbitrary numerical function of time, and also, to imagine the coupling term qQ supplied with a variable factor λ. Then

$$\begin{aligned}\frac{\partial}{\partial\lambda}\langle t_2|t_2\rangle^{F_\pm q_\pm'}&\\ &=i\left\langle\left|\int dt(q_+Q_+-q_-Q_-)\right|\right\rangle,\\ &=-i\int_{t_2}^{t_1}dt\left(\frac{\delta}{\delta F_+(t)}\frac{\delta}{\delta q_+'(t)}-\frac{\delta}{\delta F_-(t)}\frac{\delta}{\delta q_-'(t)}\right)\\ &\qquad\times\langle t_2|t_2\rangle^{F_\pm q_\pm'},\end{aligned}$$

provided that the states to which the transformation function refers do not depend upon the coupling between the systems, or that the trace of the transformation function is being evaluated. A similar statement would apply to a transformation function with different terminal times. This differential equation implies an integrated form, in which the transformation function for the fully coupled system ($\lambda=1$) is expressed in terms of the transformation function for the uncoupled system ($\lambda=0$). The latter is the product of transformation functions for the independent oscillator and external system. The relation is

$$\begin{aligned}\langle t_2|t_2\rangle^{F_\pm}=\exp\left[-i\int_{t_2}^{t_1}dt\left(\frac{\delta}{\delta F_+}\frac{\delta}{\delta q_+'}-\frac{\delta}{\delta F_-}\frac{\delta}{\delta q_-'}\right)\right]&\\ \times\langle t_2|t_2\rangle_{\text{osc}}^{F_\pm}\langle t_2|t_2\rangle_{\text{ext}}^{q_\pm'}\big|_{q_\pm'=0},&\end{aligned}$$

and we have indicated that $q_\pm'$ is finally set equal to zero if we are concerned only with measurements on the oscillator.

Let us consider for the moment just the external system with the perturbation $q'Q$, the effect of which is indicated by[6]

$$\langle t_2|t_2\rangle^{q_\pm'}=\left\langle t_2\left|\left(\exp\left[-i\int dtq_-'Q\right]\right)_-\right.\right.$$
$$\times\left.\left.\left(\exp\left[i\int dtq_+'Q\right]\right)_+\right|t_2\right\rangle.$$

We shall define

$$Q_+(t,q_\pm')=\frac{\langle t_2|Q_+(t)|t_2\rangle^{q_\pm'}}{\langle t_2|t_2\rangle^{q_\pm'}}$$
$$=\frac{1}{i}\frac{\delta}{\delta q_+'(t)}\log\langle t_2|t_2\rangle^{q_\pm'}$$

and similarly

$$Q_-(t,q_\pm')=\frac{\langle t_2|Q_-(t)|t_2\rangle^{q_\pm'}}{\langle t_2|t_2\rangle^{q_\pm'}}$$
$$=-\frac{1}{i}\frac{\delta}{\delta q_-'(t)}\log\langle t_2|t_2\rangle^{q_\pm'}.$$

When $q_\pm'(t)=q'(t)$, we have

$$Q_+(t,q')=Q_-(t,q')=\langle t_2|Q(t)|t_2\rangle^{q'},$$

which is the expectation value of $Q(t)$ in the presence of the perturbation described by $q'(t)$. This is assumed to be zero for $q'(t)=0$ and depends generally upon the history of $q'(t)$ between t_2 and the given time.

The operators $q_\pm(t)$ are produced within the transformation function by the functional differential operators $(\pm 1/i)\ \delta/\delta F_\pm(t)$, and since the equation of motion for the uncoupled oscillator is

$$\left(\frac{d^2}{dt^2}+\omega_0^2\right)q(t)=F(t),$$

we have

$$\left(\frac{\partial^2}{\partial t^2}+\omega_0^2\right)\left(\pm\frac{1}{i}\right)\frac{\delta}{\delta F_\pm(t)}\langle t_2|t_2\rangle^{F\pm}$$
$$=\exp\left[-i\int dt\left(\frac{\delta}{\delta F_+}\frac{\delta}{\delta q_+'}-\frac{\delta}{\delta F_-}\frac{\delta}{\delta q_-'}\right)\right]$$
$$\times F_\pm(t)\langle t_2|t_2\rangle_{\text{osc}}{}^{F\pm}\langle t_2|t_2\rangle_{\text{ext}}{}^{q_\pm'}|_{q_\pm'=0}.$$

On moving $F_\pm(t)$ to the left of the exponential, this becomes

$$F_\pm(t)\langle t_2|t_2\rangle^{F\pm}+\exp[\ ]\langle t_2|t_2\rangle_{\text{osc}}{}^{F\pm}$$
$$\times\left(\pm\frac{1}{i}\right)\frac{\delta}{\delta q_\pm'(t)}\langle t_2|t_2\rangle_{\text{ext}}{}^{q_\pm'}|_{q_\pm'=0}.$$

But

$$\left(\pm\frac{1}{i}\right)\frac{\delta}{\delta q_\pm'(t)}\langle t_2|t_2\rangle_{\text{ext}}{}^{q_\pm'}=Q_\pm(t,q_\pm')\langle t_2|t_2\rangle_{\text{ext}}{}^{q_\pm'},$$

and furthermore,

$$\exp[\ ]Q(t,q_\pm')\langle t_2|t_2\rangle_{\text{osc}}{}^{F\pm}\langle t_2|t_2\rangle_{\text{ext}}{}^{q_\pm'}|_{q_\pm'=0}$$
$$=Q\left(t,\pm\frac{1}{i}\frac{\delta}{\delta F_\pm}\right)\langle t_2|t_2\rangle^{F\pm},$$

which leads us to the following functional differential equation for the transformation function $\langle t_2|t_2\rangle^{F\pm}$, in which a knowledge is assumed of the external system's reaction to the perturbation $q_\pm'(t)$:

$$\left[\left(\frac{\partial^2}{\partial t^2}+\omega_0^2\right)\left(\pm\frac{1}{i}\right)\frac{\delta}{\delta F_\pm(t)}-Q_\pm\left(t,\pm\frac{1}{i}\frac{\delta}{\delta F_\pm}\right)-F_\pm(t)\right]$$
$$\times\langle t_2|t_2\rangle^{F\pm}=0.$$

Throughout this discussion one must distinguish between the $\pm$ signs attached to particular components and those involved in the listing of complete sets of variables.

The differential equations for time development are supplemented by boundary conditions which assert, at a time t_1 beyond which $F_+(t)=F_-(t)$, that

$$\left(\frac{\delta}{\delta F_+(t_1)}+\frac{\delta}{\delta F_-(t_1)}\right)\langle t_2|t_2\rangle^{F\pm}=0$$

while, for the example of the transformation function $\langle t_2|t_2\rangle_{\vartheta_0\vartheta}{}^{F\pm}$, we have the initial conditions

$$\left[\left(\frac{\delta}{\delta F_+}-\frac{\delta}{\delta F_-}\right)(t_2)\right.$$
$$\left.+\frac{i}{\omega_0}\coth(\tfrac{1}{2}\omega_0\beta_0)\frac{\partial}{\partial t_2}\left(\frac{\delta}{\delta F_+}+\frac{\delta}{\delta F_-}\right)(t_2)\right]\langle t_2|t_2\rangle^{F\pm}=0$$

and

$$\left[\frac{\partial}{\partial t_2}\left(\frac{\delta}{\delta F_+}-\frac{\delta}{\delta F_-}\right)(t_2)\right.$$
$$\left.-i\omega_0\coth(\tfrac{1}{2}\omega_0\beta_0)\left(\frac{\delta}{\delta F_+}+\frac{\delta}{\delta F_-}\right)(t_2)\right]\langle t_2|t_2\rangle^{F\pm}=0.$$

The previous treatment can now be identified as the

[6] Such positive and negative time-ordered products occur in a recent paper by K. Symanzik [J. Math. Phys. 1, 249 (1960)], which appeared after this paper had been written and its contents used as a basis for lectures delivered at the Brandeis Summer School, July, 1960.

approximation of the $Q_\pm(t,q_\pm')$ by linear functions of $q_\pm'$,

$$Q_+(t,q_\pm')=i\int dt'[A_{++}(t-t')q_+'(t')-A_{+-}(t-t')q_-'(t')]$$

$$Q_-(t,q_\pm')=i\int dt'[A_{-+}(t-t')q_+'(t')-A_{--}(t-t')q_-'(t')],$$

wherein the linear equations for the operators $q_\pm(t)$ and their meaning in terms of variations of the $F_\pm$ have been united in one pair of functional differential equations. This relation becomes clearer if one writes

$$\langle t_2|t_2\rangle^{F_\pm}=\exp[iW(F_\pm)]$$

and, with the definition

$$q_\pm(t,F_\pm)=(\pm)\frac{\delta}{\delta F_\pm(t)}W(F_\pm)$$

$$=\left(\pm\frac{1}{i}\right)\frac{\delta}{\delta F_\pm(t)}\log\langle t_2|t_2\rangle^{F_\pm},$$

converts the functional differential equations into

$$\left(\frac{\partial^2}{\partial t^2}+\omega_0^2\right)q_\pm(t,F_\pm)-Q_\pm\left(t,\,q_\pm\pm\frac{1}{i}\frac{\delta}{\delta F_\pm}\right)-F_\pm(t)=0.$$

The boundary conditions now appear as

$$(q_+-q_-)(t_1,F_\pm)=0$$

and

$$(q_++q_-)(t_2,F_\pm)+\frac{i}{\omega_0}\coth(\tfrac{1}{2}\omega_0\beta_0)\frac{\partial}{\partial t_2}(q_+-q_-)(t_2,F_\pm)=0$$

$$\frac{\partial}{\partial t_2}(q_++q_-)(t_2,F_\pm)-i\omega_0\coth(\tfrac{1}{2}\omega_0\beta_0)(q_+-q_-)(t_2,F_\pm)=0.$$

When the $Q_\pm$ are linear functions of $q_\pm$, the functional differential operators disappear[7] and we regain the linear equations for $q_\pm(t)$, which in turn imply the quadratic form of $W(F_\pm)$ that characterizes the preceding discussion.

[7] The degeneration of the functional equations into ordinary differential equations also occurs when the motion of the oscillator is classical and free of fluctuation.

9.2 COULOMB GREEN'S FUNCTION[†]

[†] Reproduced from the Journal of Mathematical Physics, Vol. 5, pp. 1606-1608 (1964).

Reprinted from *Printed in U.S.A.*
JOURNAL OF MATHEMATICAL PHYSICS VOLUME 5, NUMBER 11 NOVEMBER 1964

Coulomb Green's Function

JULIAN SCHWINGER
Harvard University, Cambridge, Massachusetts
(Received 19 June 1964)

A one-parameter integral representation is given for the momentum space Green's function of the nonrelativistic Coulomb problem.

It has long been known that the degeneracy of the bound states in the nonrelativistic Coulomb problem can be described by a four-dimensional Euclidean rotation group, and that the momentum representation is most convenient for realizing the connection. It seems not to have been recognized, however, that the same approach can be used to obtain an explicit construction for the Green's function of this problem. The derivation[1] is given here.

The momentum representation equation for the Green's function is ($\hbar = 1$)

$$\left(E - \frac{p^2}{2m}\right)G(\mathbf{p}, \mathbf{p}') + \frac{Ze^2}{2\pi^2}\int (d\mathbf{p}'')\,\frac{1}{(\mathbf{p} - \mathbf{p}'')^2} \times G(\mathbf{p}'', \mathbf{p}') = \delta(\mathbf{p} - \mathbf{p}').$$

We shall solve this equation by assuming, at first, that

$$E = -(p_0^2/2m)$$

is real and negative. The general result is inferred by analytic continuation.

The parameters

$$\xi = -\frac{2p_0\mathbf{p}}{p_0^2 + p^2}, \qquad \xi_0 = \frac{p_0^2 - p^2}{p_0^2 + p^2}$$

define the surface of a unit four-dimensional Eu-

[1] It was worked out to present at a Harvard quantum mechanics course given in the late 1940's. I have been stimulated to rescue it from the quiet death of lecture notes by recent publications in this Journal, which give alternative forms of the Green's function: E. H. Wichmann and C. H. Woo, J. Math. Phys. **2**, 178 (1961); L. Hostler, *ibid.* **5**, 591 (1964).

clidean sphere,

$$\xi_0^2 + \xi^2 = 1,$$

the points of which are in one to one correspondence with the momentum space. The element of area on the sphere is

$$d\Omega = \frac{(d\xi)}{|\xi_0|} = \left(\frac{2p_0}{p_0^2 + p^2}\right)^3 (d\mathbf{p}),$$

if one keeps in mind that $p \gtrless p_0$ corresponds to the two semispheres $\xi_0 = \mp(1 - \xi^2)^{\frac{1}{2}}$. As another form of this relation, we write the delta function connecting two points on the unit sphere as

$$\delta(\Omega - \Omega') = \left(\frac{p_0^2 + p^2}{2p_0}\right)^3 \delta(\mathbf{p} - \mathbf{p}').$$

Next, observe that

$$\begin{aligned}(\xi - \xi')^2 &= (\xi_0 - \xi_0')^2 + (\xi - \xi')^2 \\ &= \frac{4p_0^2}{(p_0^2 + p^2)(p_0^2 + p'^2)} (\mathbf{p} - \mathbf{p}')^2.\end{aligned}$$

Then, if we define

$$\Gamma(\Omega, \Omega') = -\frac{1}{16mp_0^3} (p_0^2 + p^2)^2 G(\mathbf{p}, \mathbf{p}')(p_0^2 + p'^2)^2,$$

that function obeys a four-dimensional Euclidean surface integral equation,

$$\Gamma(\Omega, \Omega') - 2\nu \int d\Omega'' \, D(\xi - \xi'')\Gamma(\Omega'', \Omega') = \delta(\Omega - \Omega'),$$

where

$$D(\xi - \xi') = \frac{1}{4\pi^2} \frac{1}{(\xi - \xi')^2}$$

and

$$\nu = Ze^2 m/p_0.$$

The function D that is defined similarly throughout the Euclidean space is the Green's function of the four-dimensional Poisson equation,

$$-\partial^2 D(\xi - \xi') = \delta(\xi - \xi').$$

It can be constructed in terms of a complete set of four-dimensional solid harmonics. In the spherical coordinates indicated by ρ, Ω, these are

$$(\rho^{n-1}, \rho^{-n-1}) Y_{nlm}(\Omega), \qquad n = 1, 2, \cdots,$$

where the quantum numbers l, m provide a three-dimensional harmonic classification of the four-dimensional harmonics. The largest value of l contained in the homogeneous polynomial $\rho^{n-1} Y_{nlm}(\Omega)$ is the degree of the polynomial, $n - 1$. Thus,

$$-l \le m \le l, \qquad 0 \le l \le n - 1$$

label the n^2 distinct harmonics that have a common value of n.

The Green's function D is exhibited as

$$D(\xi - \xi') = \sum_{n=1}^{\infty} \frac{\rho_<^{n-1}}{\rho_>^{n+1}} \frac{1}{2n} \sum_{lm} Y_{nlm}(\Omega) Y_{nlm}(\Omega')^*,$$

where

$$\delta(\Omega - \Omega') = \sum_{nlm} Y_{nlm}(\Omega) Y_{nlm}(\Omega')^*$$

conveys the normalization and completeness of the surface harmonics. One can verify that D has the radial discontinuity implied by the delta function inhomogeneity of the differential equation,

$$-\rho^3 \frac{\partial}{\partial \rho} D(\xi - \xi') \Big]_{\rho' - 0}^{\rho' + 0} = \delta(\Omega - \Omega').$$

The function D is used in the integral equation for Γ with $\rho = \rho' = 1$. The equation is solved by

$$\Gamma(\Omega, \Omega') = \sum_{nlm} \frac{Y_{nlm}(\Omega) Y_{nlm}(\Omega')^*}{1 - (\nu/n)}.$$

The singularities of this function at $\nu = n = 1, 2, \cdots$ give the expected negative energy eigenvalues. The residues of G at the corresponding poles in the E plane provide the normalized wavefunctions, which are

$$\Psi_{nlm}(p) = \frac{4p_0^{5/2}}{(p_0^2 + p^2)^2} Y_{nlm}(\Omega),$$

$$p_0 = Ze^2 m/n.$$

One can exhibit $\Gamma(\Omega, \Omega')$ in essentially closed form with the end of the expansion for D. We use the following version of this expansion:

$$\frac{1}{2\pi^2} \frac{1}{(1 - \rho)^2 + \rho(\xi - \xi')^2} = \sum_{n=1}^{\infty} \rho^{n-1} \frac{1}{n} \sum_{lm} Y_{nlm}(\Omega) Y_{nlm}(\Omega')^*,$$

where ξ and ξ' are of unit length and $0 < \rho < 1$. Note, incidentally, that if we set $\xi = \xi'$ and integrate over the unit sphere, of area $2\pi^2$, we get

$$\frac{1}{(1 - \rho)^2} = \sum_{n=1}^{\infty} \rho^{n-1} \frac{1}{n} m_n,$$

where m_n is the multiplicity of the quantum number n. This confirms that $m_n = n^2$.

The identity

$$\frac{1}{1 - (\nu/n)} = 1 + \frac{\nu}{n} + \nu^2 \frac{1}{n(n - \nu)},$$

together with the integral representation

$$\frac{1}{n - \nu} = \int_0^1 d\rho \rho^{-\nu} \rho^{n-1},$$

valid for $\nu < 1$, gives

$$\Gamma(\Omega, \Omega') = \delta(\Omega - \Omega') + \frac{\nu}{2\pi^2}\frac{1}{(\xi - \xi')^2} + \frac{\nu^2}{2\pi^2}\int_0^1 d\rho\rho^{-\nu}\frac{1}{(1-\rho)^2 + \rho(\xi - \xi')^2}. \quad (1)$$

Equivalent forms, produced by partial integrations, are

$$\Gamma(\Omega, \Omega') = \delta(\Omega - \Omega') + \frac{\nu}{2\pi^2}\int_0^1 d\rho\rho^{-\nu}\frac{d}{d\rho}\frac{\rho}{(1-\rho)^2 + \rho(\xi - \xi')^2}, \quad (2)$$

and

$$\Gamma(\Omega, \Omega') = \frac{1}{2\pi^2}\int_0^1 d\rho\rho^{-\nu}\frac{d}{d\rho}\frac{\rho(1-\rho^2)}{[(1-\rho)^2 + \rho(\xi - \xi')^2]^2}, \quad (3)$$

which uses the limiting relation

$$\delta(\Omega - \Omega') = \lim_{\rho\to 1}\frac{1}{2\pi^2}\frac{1-\rho^2}{[(1-\rho)^2 + \rho(\xi - \xi')^2]^2}.$$

Note that Γ is a function of a single variable, $(\xi - \xi')^2$.

The restriction $\nu < 1$ can be removed by replacing the real integrals with contour integrals,

$$\int_0^1 d\rho\rho^{-\nu}(\ ) \to \frac{i}{2\sin\pi\nu}e^{\pi i\nu}\int_C d\rho\rho^{-\nu}(\ ).$$

The path C begins at $\rho = 1 + 0i$, where the phase of ρ is zero and terminates at $\rho = 1 - 0i$, after encircling the origin within the unit circle.

The Green's function expressions implied by (1), (2), and (3) are

$$G(\mathbf{p}, \mathbf{p}') = \frac{\delta(\mathbf{p} - \mathbf{p}')}{E - T} - \frac{Ze^2}{2\pi^2}\frac{1}{E-T}\frac{1}{(\mathbf{p}-\mathbf{p}')^2}\frac{1}{E-T'} - \frac{Ze^2}{2\pi^2}\frac{1}{E-T}\left[i\eta\int_0^1 d\rho\rho^{-i\eta} \times \frac{1}{(\mathbf{p}-\mathbf{p}')^2\rho - (m/2E)(E-T)(E-T')(1-\rho)^2}\right]\frac{1}{E-T'}, \quad (1')$$

where

$$T = p^2/2m, \qquad \eta = -i\nu = Ze^2m/k;$$

$$G(\mathbf{p}, \mathbf{p}') = \frac{\delta(\mathbf{p} - \mathbf{p}')}{E - T} - \frac{Ze^2}{2\pi^2}\frac{1}{E-T}\left[\int_0^1 d\rho\rho^{-i\eta}\frac{d}{d\rho} \times \frac{\rho}{(\mathbf{p}-\mathbf{p}')^2\rho - (m/2E)(E-T)(E-T')(1-\rho)^2}\right]\frac{1}{E-T'}; \quad (2')$$

and

$$G(\mathbf{p}, \mathbf{p}') = -\frac{i}{4\pi^2}\frac{k}{E}\int_0^1 d\rho\rho^{-i\eta}\frac{d}{d\rho} \times \frac{\rho(1-\rho^2)}{[(\mathbf{p}-\mathbf{p}')^2\rho - (m/2E)(E-T)(E-T')(1-\rho)^2]^2}. \quad (3')$$

The Green's function is regular everywhere in the complex E plane with the exception of the physical energy spectrum. This consists of the negative-energy eigenvalues already identified and the positive-energy continuum. The integral representations (1′), (2′), and (3′) are not completely general since it is required that

$$\operatorname{Re} i\eta = -\operatorname{Im}\eta < 1.$$

As we have indicated, this restriction can be removed. It is not necessary to do so, however, if one is interested in the limit of real k. These representations can therefore be applied directly to the physical scattering problem.

The asymptotic conditions that characterize finite angle deflections are

$$E - T \sim 0, \qquad E - T' \sim 0, \qquad (\mathbf{p} - \mathbf{p}')^2 > 0.$$

The second of the three forms given for G is most convenient here. The asymptotic behavior is dominated by small ρ values, and one immediately obtains

$$G(\mathbf{p}, \mathbf{p}') \sim G^0(\mathbf{p})(-1/4\pi^2 m)f(\mathbf{p}, \mathbf{p}')G^0(\mathbf{p}'),$$

where

$$G^0(\mathbf{p}) = \frac{1}{E-T}\exp\left[-i\eta\log\frac{E-T}{4E}\right]\left(\frac{2\pi\eta}{e^{2\pi\eta}-1}\right)^{\frac{1}{2}}$$

and

$$f(\mathbf{p}, \mathbf{p}') = \frac{2mZe^2}{(\mathbf{p}-\mathbf{p}')^2}\exp\left[-i\eta\log\frac{4k^2}{(\mathbf{p}-\mathbf{p}')^2}\right],$$

$$p^2 = p'^2 = k^2.$$

One would have found the same asymptotic form for any potential that decreases more rapidly than the Coulomb potential at large distances, but with $G^0(\mathbf{p}) = (E - T)^{-1}$. The factors $G^0(\mathbf{p}')$ and $G^0(\mathbf{p})$ describe the propagation of the particle before and after the collision, respectively, and f is identified as the scattering amplitude. The same interpretation is applicable here since the modified G^0 just incorporates the long-range effect of the Coulomb potential. This is most evident from the asymptotic behavior of the corresponding spatial function, which is a distorted spherical wave,

$$\int\frac{(d\mathbf{p})}{(2\pi)^3}e^{i\mathbf{p}\cdot\mathbf{r}}G(\mathbf{p}) \sim (-m/2\pi r)\exp[i(kr + \eta\log 2kr + \zeta)],$$

$$\zeta = \arg\Gamma(1 - i\eta).$$

The scattering amplitude obtained in this way coincides with the known result,

$$f(\vartheta) = (Ze^2/4E)\csc^2\tfrac{1}{2}\vartheta\exp[-i\eta\log\csc^2\tfrac{1}{2}\vartheta].$$